Multiple Choice Questions in Biochemistry

H HASSALL BSc, PhD
A J TURNER MA, PhD
E J WOOD MA, DPhil

Department of Biochemistry
University of Leeds

PITMAN

PITMAN PUBLISHING LIMITED
128 Long Acre, London WC2E 9AN

Associated Companies

Pitman Publishing Pty Ltd, Melbourne
Pitman Publishing New Zealand Ltd, Wellington

Co-published by Urban & Schwarzenberg, Baltimore

First published 1985

Library of Congress Cataloging in Publication Data

Hassall, H.
    Multiple choice questions in biochemistry.

    1. Biological chemistry—Examinations, questions, etc.
I. Turner, A. J. II. Wood, E. J. III. Title.
[DNLM: 1. Biochemistry—examination questions.
QU 18 H353m]
QP518.5.H37 1984   574.19′2′076   84-14877
ISBN 0-272-79758-8 (pbk.)

British Library Cataloguing in Publication Data

Hassall, H.
    Multiple choice questions in biochemistry.
    1. Biochemistry—Problems, exercises, etc.
    I. Title II. Turner, A.J. III. Wood, E. J.
    574.19′2′076   QP518.5

    ISBN 0-272-79758-8

Printed and bound in Great Britain at The Pitman Press, Bath

Contents

Foreword

There is much to be said for the view that examinations are a 'necessary evil' in that it is a nuisance for the teacher to have to impose them on an otherwise excellent academic syllabus. It is probably worth making this point, since students tend to think that teachers revel in exams and that they enjoy the spectacle of failure. Nothing could be further from the truth in the Universities in which I have taught. We would be delighted to see all students pass first time.

The MCQ test is a comparatively new form of examination. There is no doubt that it gained acceptance during a period of rapid growth in education. MCQ tests easily lend themselves to computer marking. Thus the mystique of the computer in contemporary life has helped the acceptance of the MCQ which is now seen as an examination free from personal bias and one that is completely objective.

The critics rightly asked what the MCQ test was assessing. Much of the informed debate took place in the USA and it now seems to be agreed that MCQ tests are useful for assessing a student's knowledge of information. They are not necessarily a way of effectively evaluating the comparative quality of a group of students in respect of other sets of criteria. For this, one may need a whole range of assessments which could include projects, problems, short answers, essays and oral examinations. Neither students nor teachers should accept ranking in an MCQ-based examination as the only indicator of an order of merit.

The MCQ test also has considerable value in the assessment of the progress of a student during a course. For this purpose, it may be more acceptable if the marks are confidential to the student.

It is not easy to produce a good MCQ examination. By this, I mean that it is difficult to write questions that are discriminatory in respect of determining whether a student has acquired the appropriate information. There are problems of ensuring technical accuracy as well as relevance to course content. The basis for students selecting answers may be completely unrelated to sound biochemical knowledge and may relate to some facet of the questions such as the relationship of the length of the statements to the likelihood of their being correct. In short, setting a good MCQ exam requires a group of teachers working together and arguing the pros and cons of each question. It is not a job for a lone teacher.

It is a pleasure to welcome this book of multiple-choice questions put together by my former colleagues at the University of Leeds. They have had experience over many years of setting similar questions for students of mixed ability. I am sure that students will find the present questions useful in assessing their progress and as practice for formal examinations. Teachers should also find the questions useful when framing their own exam papers, although they will no doubt wish to adapt them to their own particular courses.

P N Campbell

Courtauld Institute of Biochemistry
Middlesex Hospital Medical School
London

Preface

The majority of students these days are subjected to multiple-choice testing from school to university. Many of them perform well, having mastered the technique. In practice, this means interpreting the questions carefully as well as having the information stored away. Others dislike this form of testing and perform badly, giving reasons such as: 'I never do well on MCQ tests', or: 'The questions are ambiguous', or even: 'Penalizing for incorrect choices has a detrimental psychological effect'. Certain criticisms may be justified and questions should always be modified in the light of comments. One might also add the charge that MCQ tests do not encourage the development by students of literacy nor the ability to synthesize information. This form of testing should therefore not stand alone in the examination of students. Nevertheless, multiple-choice testing is here to stay and there are a variety of reasons for this, many of them economic rather than academic. It also becomes relatively easy to examine large classes, especially using computer-marking, and a certain degree of standardization is feasible.

A major reason for writing this book was to present a set of questions enabling students to test themselves, providing more detailed answers than is common in books of this type. We hope, therefore, that the book will bridge the gap between the straightforward collection of test questions, which provide minimal explanatory information, and the standard textbook. We have tried to be topical and to include a wide range of subject matter, from medical biochemistry through to plant biochemistry. We therefore hope that aspects of the book will appeal to a wide range of students.

The questions themselves are arranged into ten tests of twenty questions each, roughly in order of increasing difficulty. Each test is followed by a summary of the answers, for quick checking. The final section of the book consisting of detailed answers to each of the questions is arranged into sections corresponding to subject matter. There is necessarily some overlap, but this should aid revision and cross-referencing. For the most part, we have adopted as a format for the questions a stem phrase followed by five possible options that complete the sentence. Any one or more of the options may be correct, and each should be considered independently. This is certainly the most common test format, particularly in medical examinations. However, it is by no means sacrosanct, and some variations of this theme have been used. In composing the questions we have haggled amongst ourselves as to what was sensible to ask, what was ambiguous and what it was reasonable to expect the student to remember. This was a valuable, if salutary, experience for us and we hope that the format, style, range and depth of questions and answers will be of assistance to others when they are writing questions. The present questions are probably a long way from being perfect but they are certainly much improved by having been considered by three, and sometimes more, people. Our final message, therefore, to question writers is: 'Always get colleagues to criticise the questions before presenting them to students' and our message to students is: 'Even if you don't like doing MCQ tests, they are almost certain to form an important part of your assessment these days and you will certainly improve with practice'.

H Hassall
A J Turner
E J Wood

Leeds, May 1984

Part 1: Random Questions

Each group of 20 questions (which are numbered as they appear in Part 2) is followed by the answers.

1.4 Mitochondria:

1. are found in all prokaryotic cells, except photosynthetic bacteria.
2. are found in all eukaryotic cells.
3. contain DNA.
4. contain ribosomes.
5. in evolutionary terms, may have originated from primitive bacteria.

2.4 The figure shows the result of an electrophoresis experiment conducted at pH 6.5 with the amino acids glycine, lysine and glutamic acid. After the electrophoresis, the paper strip was dried and stained with ninhydrin to reveal the positions of the amino acids.

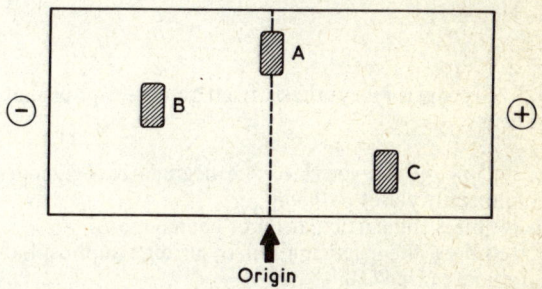

Which of the following deductions is true?

1. Amino acid C is glycine.
2. Amino acid B is lysine.
3. Amino acid A is glutamic acid.
4. Amino acid A has an isoelectric point near pH 6.5.
5. At pH 8, all the amino acids, A, B, and C, would move towards the positive electrode.

3.1 As applied to a protein, the term primary structure refers to:

1. The *number* of amino acid residues present in the protein.

2. the total amino acid *composition* of the protein.
3. the *sequence of amino acids* in the protein.
4. the α-*helical region* of the protein.
5. the *geometrical organization* of the polypeptide chain.

4.9 The initial rate of an enzyme-catalysed reaction is:

1. normally directly proportional to substrate concentration.
2. normally directly proportional to enzyme concentration.
3. always maximal at neutral pH.
4. always maximal when the substrate concentration is equal to the K_m (Michaelis constant).
5. independent of temperature.

5.3 'Reducing' sugars include:

1. glucose.
2. fructose.
3. sucrose.
4. galactose.
5. lactose.

6.2 Glycogen biosynthesis from glucose 6-phosphate in liver:

1. occurs by a reversal of the degradative (glycogen phosphorylase) pathway.
2. requires the participation of coenzyme A.
3. requires the participation of uridine diphosphate glucose (UDPG).
4. requires the participation of a branching enzyme.
5. is an energy-requiring reaction.

6.12 Gluconeogenesis:

1. is the formation of glucose from glycogen.
2. takes place primarily in the liver.
3. may use acetyl CoA as a precursor.
4. is stimulated during starvation.
5. allows the resynthesis of glucose and glycogen from lactate after vigorous exercise.

7.4 Catabolic pathways may operate to:

1. use up excess ATP.
2. provide reducing power (as NADPH).
3. provide energy (for example as ATP).
4. keep the organism warm.
5. provide low molecular weight compounds which serve as precursors for biosynthetic pathways.

8.7 The process of fatty acid biosynthesis shows several features which distinguish it clearly from fatty acid oxidation. These include:

1. its occurrence in the cytosol instead of in the mitochondria.
2. the requirement for ferredoxin.
3. the requirement for NADPH.
4. the participation of biotin.
5. the requirement for $FADH_2$.

9.4 The urea cycle:

1. supplies the bodily requirement for arginine in infants.
2. converts urea into uric acid.
3. converts ammonia into urea.
4. acts as an energy-supplying mechanism by oxidizing waste materials.
5. converts urea to ammonia and carbon dioxide.

10.5 Fructose 1,6-bisphosphatase:

1. is a key regulatory enzyme in glycolysis.
2. is activated by ATP.
3. is activated by citrate.
4. is activated by cyclic AMP-dependent phosphorylation.
5. is inhibited by AMP.

12.1 In nucleic acids:

1. the normal nitrogenous bases in RNA are adenine, guanine, cytosine and uracil.
2. the bases in DNA are adenine, guanine, cytosine and thiamine.
3. the total purine content of native DNA is always equal to the total pyrimidine content.

4. the total purine content of RNA is always equal to the total pyrimidine content.
5. information for protein synthesis is stored as a triplet code in DNA.

13.6 Protein biosynthesis in both eukaryotes and prokaryotes:

1. begins with the N-terminus of the polypeptide.
2. occurs on 80S ribosomes.
3. is inhibited by puromycin.
4. is inhibited by cycloheximide.
5. uses AUG as the initiation codon.

17.1 The immunoglobulin G molecule (IgG) is composed of two pairs of polypeptides, two 'heavy' and two 'light'. Each 'heavy' chain:

1. has a M_r of about 150 000.
2. consists of four 'domains'.
3. is 'variable' in amino acid sequence for about half its length.
4. is attached to a light chain and to the other heavy chain by disulphide bonds.
5. contains covalently bound carbohydrate.

18.7 Available energy for muscle contraction is obtained during anaerobic glycolysis from the following individual steps:

1. the formation of fructose 1,6-bisphosphate from fructose 6-phosphate.
2. the conversion of 3-phosphoglyceraldehyde to 1,3-glycerate bisphosphate (1,3-diphosphoglycerate).
3. the conversion of 1,3-diphosphoglycerate to 3-phosphoglycerate.
4. the conversion of phosphoenolpyruvate to pyruvate.
5. the reduction of pyruvate to lactate.

19.1 For nutritional well-being, the mammalian diet must contain, in addition to vitamins, certain specific:

1. amino acids.
2. fatty acids.
3. purines and pyrimidines.

4. monosaccharides.
5. metal ions.

20.3 Galactosaemia:

1. is characterized by an accumulation of galactose and galactose 1-phosphate in blood and other tissues.
2. most commonly results from a deficiency in galactokinase.
3. is associated with mental deficiency and can result in impaired vision.
4. is a genetically linked disease.
5. is less critical in adults than in children.

21.7 Proteins are commonly purified by:

1. ion-exchange chromatography.
2. paper chromatography.
3. gel filtration (gel exclusion) chromatography.
4. salt precipitation.
5. extraction with phenol.

21.9 Pancreatic juice:

1. has a pH of approximately 7 to 8.
2. contains bile salts.
3. contains insulin.
4. contains trypsinogen.
5. contains α-1,4-glucosidase (acid maltase).

22.3 In the diagram of chloroplast, shown below:

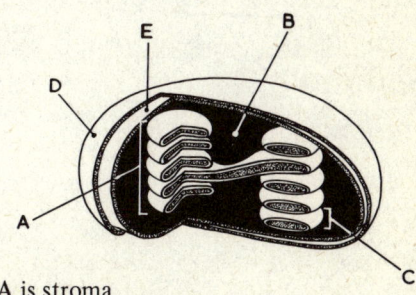

1. A is stroma.
2. B is granum.
3. C is a thylakoid.
4. D is outer membrane.
5. E is endoplasmic reticulum.

Question Number		Answers			
1.4	1.F	2.F	3.T	4.T	5.T
2.4	1.F	2.T	3.F	4.T	5.F
3.1	1.F	2.F	3.T	4.F	5.F
4.9	1.F	2.T	3.F	4.F	5.F
5.3	1.T	2.T	3.F	4.T	5.T
6.2	1.F	2.F	3.T	4.T	5.T
6.12	1.F	2.T	3.F	4.T	5.T
7.4	1.F	2.T	3.T	4.T	5.T
8.7	1.T	2.F	3.T	4.T	5.F
9.4	1.F	2.F	3.T	4.F	5.F
10.5	1.F	2.T	3.T	4.F	5.T
12.1	1.T	2.F	3.T	4.F	5.T
13.6	1.T	2.F	3.T	4.F	5.T
17.1	1.F	2.T	3.F	4.T	5.T
18.7	1.F	2.F	3.T	4.T	5.F
19.1	1.T	2.T	3.F	4.F	5.T
20.3	1.T	2.F	3.T	4.T	5.T
21.7	1.T	2.F	3.T	4.T	5.F
21.9	1.T	2.F	3.F	4.T	5.F
22.3	1.F	2.F	3.T	4.T	5.F

1.9 The eukaryotic (80S) ribosome:

1. dissociates into subunits with sedimentation coefficients of 50S and 30S.
2. is a multi-enzyme complex composed of RNA and protein.
3. contains two binding sites for transfer RNA (tRNA) molecules.
4. is primarily assembled in the nucleus.
5. dissociates into its constituent subunits on completion of protein synthesis.

2.3 The graph shows the titration curve for glycine hydrochloride.

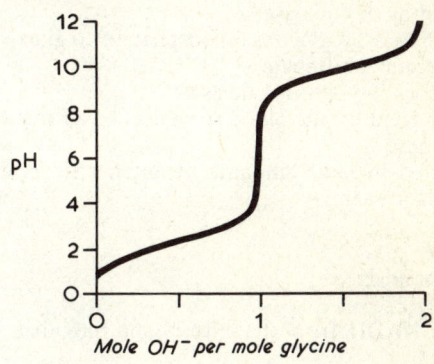

Mole OH^- per mole glycine

Glycine:

1. would act as a buffer over the pH range 1.3 to 3.3.
2. would act as a buffer over the pH range 4.3 to 6.3.
3. would act as a buffer over the pH range 8.6 to 10.6.
4. would carry zero net charge at pH 6.0.
5. would, at pH 2.3, exist in equimolar amounts of $^+H_3NCH_2CO_2H$ and $^+H_3NCH_2CO_2^-$.

4.6 The Michaelis-Menten model of enzyme action:

1. assumes formation of an enzyme–substrate complex.
2. assumes formation of a covalent intermediate between enzyme and substrate.
3. explains the allosteric behaviour of certain regulatory enzymes.
4. explains the stereospecificity of enzyme reactions.
5. explains the maximum rate (V_{max}) attainable in enzyme-catalysed reactions.

5.4 Carbohydrate is stored as:

1. polyglucans in most living organisms.
2. cellulose in green plants.
3. lactose in the mammary gland.
4. glycogen in mammalian muscle as a source of blood glucose.
5. glycoproteins.

6.8 Glucose 6-phosphate:

1. is formed from glucose and ATP in the reaction catalysed by hexokinase.
2. is hydrolysed by glucose 6-phosphatase to glucose and inorganic phosphate.
3. is a 'high energy' phosphate ester.
4. is transported by the blood from liver to all tissues of the body.
5. is excreted in large amounts by infants suffering from galactosaemia.

7.5 Mitochondria:

1. take up NADH from the cytosol and reoxidize it to NAD^+.
2. transport one molecule of ATP to the cytosol for every molecule of ADP taken up.
3. contain all of the succinate dehydrogenase present in a cell.
4. are freely permeable to NH_3 but not to NH_4^+ ions.
5. contain high levels of glucose 6-phosphate dehydrogenase.

8.3 The breakdown of the fatty acid

$$CH_3 CH_2 CH_2 CH_2 CH_2 CH_2 CH_2 CH_2 CO_2 H$$

via the β-oxidation pathway would:

1. not occur unless the coenzyme A derivative was formed first.
2. yield four mol acetyl units only.
3. yield four mol acetyl units and one mol propionyl units.
4. yield three mol propionyl units.
5. yield three mol acetyl units and one mol propionyl units.

9.6   Transamination (aminotransferase) reactions:

1. require the presence of coenzyme A.
2. convert one keto-acid to an amino acid whilst simultaneously converting another amino acid to its corresponding keto-acid.
3. involve the intermediate formation of a Schiff base.
4. will not operate properly in a vitamin B_6-deficient animal.
5. involve ATP hydrolysis to give AMP and pyrophosphate (PP_i).

10.8   The change in the rate of enzyme synthesis occurring in many bacteria in response to changes in the nutritional environment:

1. is known as allosteric regulation.
2. is known as induction or repression.
3. is controlled at the level of translation.
4. frequently involves polycistronic messenger RNA.
5. is a mechanism which minimizes wasteful protein synthesis.

11.3 Adenosine monophosphate (AMP):

1. is a nucleoside.
2. can be cyclized in a reaction catalysed by adenylate cyclase to give cyclic 3′,5′-AMP.
3. is a component of both RNA and DNA.
4. is an allosteric activator of phosphofructokinase.
5. is formed, with ATP, from two molecules of ADP by the action of myokinase (adenylate kinase).

12.2   The two complementary strands of double-stranded DNA:

1. are held together by hydrophobic interactions.
2. are held together by hydrogen bonding.
3. will have the same amount of any one of the four bases in each strand.
4. may be separated from each other by gently heating in solution.
5. are held together with different strengths depending upon the source of the DNA.

13.1 Several types of ribonucleic acid (RNA) can be found in cells. Of these:

1. ribosomal RNA (rRNA) is synthesized in the nucleolus.
2. messenger RNA (mRNA) is synthesized in the nucleus and carries the genetic information to the cytoplasm.
3. messenger RNA (mRNA) is the major type of RNA found in the cell.
4. transfer RNA (tRNA) acts as an 'adaptor' between the amino acid and the triplet code.
5. transfer RNA (tRNA) exists as at least 20 different kinds.

14.3 A phosphoglyceride such as lecithin is made up from glycerol, phosphate, and:

1. one molecule of long chain fatty acid and choline.
2. two molecules of long chain fatty acid and choline.
3. three molecules of long chain fatty acid and choline.
4. three molecules of long chain fatty acid and ethanolamine.
5. three molecules of long chain fatty acid and inositol.

15.2 Membrane proteins:

1. are synthesized on free ribosomes.
2. cannot be removed from the membrane in an active form.
3. can diffuse in the plane of the lipid bilayer.
4. are arranged symmetrically across the lipid bilayer.
5. may comprise 50% or more of the mass of the membrane.

16.2 Insulin:

1. is synthesized in the β-cells of the pancreatic islets.
2. is composed of two polypeptide chains.
3. is secreted in response to a fall in plasma glucose concentration.
4. activates glycogen phosphorylase in liver.
5. is absent in maturity-onset diabetes.

17.3 Several classes of immunoglobulin molecule are found in the human. Which of the following is true?

1. IgM is found only in the blood.
2. IgA is found only in the blood.
3. IgE is found in association with mast cells in the skin.
4. Only IgG can cross the placental barrier.

5. IgG is the first type of humoral antibody to appear in the serum in response to antigenic challenge.

19.3 Ascorbic acid (vitamin C):

1. is a heat-stable, fat-soluble vitamin.
2. is required as a cofactor for the hydroxylation of certain proteins.
3. if deficient in the diet eventually leads to sub-cutaneous haemorrhaging, easy bruising, and bleeding and painful gums.
4. is excreted in the urine mainly unchanged if ingested in large amounts.
5. is present in high concentrations in cod-liver oil.

20.4 The urinary excretion of large amounts of phenylalanine and phenylpyruvate, occurring in phenylketonuria, results from:

1. a deficiency in tyrosine transaminase (amino-transferase).
2. a deficiency in phenylalanine hydroxylase.
3. a genetic defect.
4. pyridoxal phosphate deficiency.
5. protein malnutrition.

21.6 Flavin adenine dinucleotide (FAD):

1. is a coenzyme for hydrogen transfer reactions.
2. contains two vitamins, riboflavin (B_2) and nico-tinamide.
3. shows an increase in absorption at 340nm when reduced.
4. is bright yellow in colour when in the oxidized form.
5. is a coenzyme for succinate dehydrogenase to which it is tightly bound.

22.9 Plants use chlorophylls during photosynthesis:

1. as light-absorbing screens to prevent damage to the sensitive photosynthetic apparatus in the thyl-akoids.
2. as hydrogen donors in the production of NADPH.
3. to make the thylakoid membrane more permeable so that a gradient of protons can be generated.
4. as photo-oxidizable compounds capable of supply-ing electrons to ferredoxin and eventually to $NADP^+$.
5. as acceptors for electrons derived from CO_2.

Question Number		Answers			
1.9	1.F	2.T	3.T	4.T	5.T
2.3	1.T	2.F	3.T	4.T	5.T
4.6	1.T	2.F	3.F	4.F	5.T
5.4	1.T	2.F	3.F	4.F	5.F
6.8	1.T	2.T	3.F	4.F	5.F
7.5	1.F	2.T	3.T	4.T	5.F
8.3	1.T	2.F	3.F	4.F	5.T
9.6	1.F	2.T	3.T	4.T	5.F
10.8	1.F	2.T	3.F	4.T	5.T
11.3	1.F	2.F	3.F	4.T	5.T
12.2	1.F	2.T	3.F	4.T	5.T
13.1	1.T	2.T	3.F	4.T	5.T
14.3	1.F	2.T	3.F	4.F	5.F
15.2	1.F	2.F	3.T	4.F	5.T
16.2	1.T	2.T	3.F	4.F	5.F
17.3	1.T	2.F	3.T	4.T	5.F
19.3	1.F	2.T	3.T	4.T	5.F
20.4	1.F	2.T	3.T	4.F	5.F
21.6	1.T	2.F	3.F	4.T	5.T
22.9	1.F	2.F	3.F	4.T	5.F

1.3 The diagram shows a cross-section of a mitochondrion. Identify the structures labelled with letters:

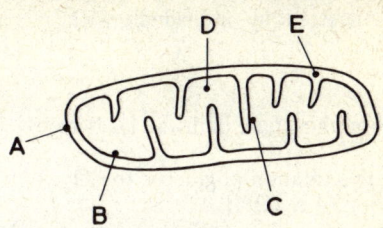

1. Structure A is the cell wall.
2. Structure B is the inner membrane.
3. Structure C is a chloroplast.
4. Structure D is 'matrix'.
5. Structure E is the space where the enzymes of oxidative phosphorylation are found.

2.7 Consider the peptide:

Ala-Gly-Ser-Pro-Tyr-Lys-Met-Ala-Lys

This peptide was treated with dansyl chloride and then hydrolysed with 6M HCl at 110°C for four hours. Examination of the products of hydrolysis by thin-layer chromatography would be expected to reveal the presence of:

1. dansyl alanine.
2. monodansyl-lysine (N-ϵ-dansyl lysine).
3. bis-dansyl-lysine (N-α,ϵ-didansyl lysine).
4. O-dansyl serine.
5. O-dansyl tyrosine.

4.5 Lactate dehydrogenase from skeletal muscle is:

1. the rate-limiting enzyme in glycolysis.
2. located in the cytosol.
3. composed of four subunits.
4. an allosteric enzyme.
5. identical in subunit composition with lactate dehydrogenase from cardiac muscle.

5.1 Lactose:

1. is galactosyl-glucose.
2. is a non-reducing sugar.

3. is incapable of being digested by many human adults.
4. shows mutarotation.
5. cannot be digested by individuals with galactosaemia.

6.1 The pentose phosphate pathway leads to:

1. the complete oxidation of glucose to CO_2.
2. the generation of NADPH.
3. the production of glucose 1-phosphate for glycogen synthesis.
4. the production of free glycerol for triglyceride synthesis.
5. the production of 5-carbon sugars (e.g. ribose) for nucleic acid synthesis.

7.1 In the Krebs tricarboxylic acid cycle (or citric acid cycle):

1. four-carbon acids such as fumarate have a catalytic effect on the rate of oxidation of pyruvate via the cycle.
2. the addition of malonate brings the cycle to a halt.
3. the addition of fluoroacetate brings the cycle to a halt.
4. 38 molecules of ATP are produced for each molecule of acetyl CoA fed into the cycle.
5. four molecules of NADH are produced for each molecule of acetyl CoA fed into the cycle.

8.9 The sequence of reactions below occurs, in the appropriate direction, in:

1. fatty acid biosynthesis.
2. fatty acid oxidation.
3. the TCA cycle.
4. glycolysis.
5. the pentose phosphate pathway.

$$-CH_2-CH_2- \overset{a}{\longleftrightarrow} -CH{=}CH- \overset{b}{\longleftrightarrow} -CHOH{-}CH_2- \overset{c}{\longleftrightarrow} -CO{-}CH_2-$$

9.2 For mammals, some amino acids are essential in the diet, whereas others may be formed from dietary components. Humans are capable of converting:

1. arginine into lysine.
2. phenylalanine into tyrosine.

3. pyruvate into alanine.
4. aspartic acid into isoleucine.
5. oxaloacetate into aspartic acid.

10.9 Phosphorylase kinase:

1. is converted to a catalytically active form on binding cyclic AMP.
2. is activated by phosphorylation of a specific tyrosine residue in the protein.
3. is activated by Ca^{2+} ions.
4. contains calmodulin as one of its constituent subunits.
5. activates glycogen phosphorylase by phosphorylation of a specific serine residue.

11.6 In humans, uric acid is:

1. the main end-product of purine metabolism.
2. present in excessive amounts in gout.
3. normally oxidized to form xanthine.
4. converted to urea in the liver.
5. excreted in larger amounts on a low-protein diet than during starvation.

12.3 Transfer RNA (tRNA):

1. is responsible for transferring mRNA from the nucleus to the cytoplasm.
2. is usually about 150 nucleotides long.
3. is single-stranded.
4. contains many uncommon or modified bases.
5. is of many different types but always has the sequence CCA at the $3'$ terminus.

13.10 The replication of DNA:

1. occurs by a 'semi-conservative' mechanism.
2. requires the presence of all four deoxyribonucleoside $5'$-diphosphates.
3. occurs only in the $5' \rightarrow 3'$ direction.
4. requires the action of DNA ligase.
5. requires a primer of RNA.

15.4 The Na$^+$/K$^+$ ATPase:

1. transports 2 Na$^+$ and 2 K$^+$ ions for each molecule of ATP hydrolysed.
2. is phosphorylated on an aspartyl residue in the course of the reaction.
3. is inhibited by digitalis glycosides.
4. is inhibited by valinomycin.
5. provides the energy, through ATP hydrolysis, for the active transport of glucose into *Escherichia coli*.

16.5 Glucagon:

1. is synthesized in the α-cells of the pancreatic islets.
2. is secreted in response to a fall in plasma glucose concentration.
3. is composed of two polypeptide chains.
4. stimulates cyclic AMP production in liver and adipose tissue.
5. promotes gluconeogenesis in liver.

17.2 The immunoglobulin molecule (IgG) is composed of two pairs of polypeptides. Each 'light' chain:

1. has a M_r of about 50 000.
2. will be a κ or a λ type depending only on its origin and not on the type of antigenic specificity.
3. contains two 'domains'.
4. would be 'variable' in amino acid sequence for approximately half its length.
5. would be 'constant' in amino acid sequence for its C-terminal half.

18.1 Myosin:

1. comprises the 'thick' filaments of the myofibril.
2. can catalyse the hydrolysis of ATP.
3. exists predominantly as a triple helical structure.
4. associates with actin.
5. is hydrolysed by trypsin to form tropomyosin.

19.6 A deficiency of thiamine (vitamin B_1):

1. causes kwashiorkor.
2. is characterized by a roughening and darkening of the skin, a condition known as pellagra.
3. results in peripheral neuritis, muscular weakness and ultimately death.
4. gives rise to an elevated serum pyruvate level.
5. is frequently associated with chronic alcoholism.

20.5 A positive test for reducing sugar in the urine:

1. is characteristic of diabetes mellitus.
2. is characteristic of lactose intolerance associated with impaired digestion of lactose.
3. may be due to the presence of certain pentoses.
4. is characteristic of fructosuria caused by a deficiency of fructokinase.
5. may be due to galactose and indicate galactosaemia.

21.3 The pigment, melanin:

1. is formed from tryptophan.
2. is formed from tyrosine.
3. is not formed in albinism.
4. is present in skin in order to absorb UV radiation to protect underlying structures.
5. always contains sulphur derived from cysteine.

22.2 In photosynthesis in green plants, represented by the overall equation

$$6 CO_2 + 12 H_2O \rightarrow C_6H_{12}O_6 + 6 O_2 + 6 H_2O$$

1. all the oxygen (O_2) evolved comes from carbon dioxide (CO_2).
2. an input of energy is required.
3. an input of 'reducing power' is required.
4. the initial product from the fixation of CO_2 is a 5-carbon sugar.
5. light of two wavelengths is required for optimal efficiency of carbon dioxide fixation.

Question Number		Answers			
1.3	1.F	2.T	3.F	4.T	5.F
2.7	1.T	2.T	3.F	4.F	5.T
4.5	1.F	2.T	3.T	4.F	5.F
5.1	1.T	2.F	3.T	4.T	5.F
6.1	1.T	2.T	3.F	4.F	5.T
7.1	1.T	2.T	3.T	4.F	5.F
8.9	1.T	2.T	3.T	4.F	5.F
9.2	1.F	2.T	3.T	4.F	5.T
10.9	1.F	2.F	3.T	4.T	5.T
11.6	1.T	2.T	3.F	4.F	5.F
12.3	1.F	2.F	3.T	4.T	5.T
13.10	1.T	2.F	3.T	4.T	5.T
15.4	1.F	2.T	3.T	4.F	5.F
16.5	1.T	2.T	3.F	4.T	5.T
17.2	1.F	2.T	3.T	4.T	5.T
18.1	1.T	2.T	3.F	4.T	5.F
19.6	1.F	2.F	3.T	4.T	5.T
20.5	1.T	2.F	3.T	4.T	5.T
21.3	1.F	2.T	3.T	4.T	5.F
22.2	1.F	2.T	3.T	4.F	5.T

1.5 The endoplasmic reticulum:

1. forms many folds and convolutions within the space inside mitochondria.
2. contains spaces, or cisternae, which serve as channels for the transport of various products through the cell, usually to its exterior.
3. may be either 'rough' or 'smooth', the former having its surface studded with ribosomes.
4. has as its major function the synthesis of carbohydrates such as glycogen or starch.
5. has a role in lipid biosynthesis.

2.1 The amino acid of structure

$$H_2N \; CHCO_2H$$
$$|$$
$$CH_2$$

1. has only two ionizable groups.
2. would absorb light in the UV region near 280 nm.
3. has a hydrophilic side chain (R^-) group.
4. can act as a precursor of adrenaline (epinephrine).
5. is essential in the diet of humans.

3.3 Two proteins, haemoglobin and lysozyme, have the following relative molecular masses (M_r values): 68 000 and 14 000, and the following isoelectric points: pH 6.8 and pH 11.0. A mixture of these two proteins could be separated by:

1. ion-exchange chromatography.
2. gel filtration.
3. dialysis.
4. electrophoresis at pH 7.0.
5. lyophilization.

4.1 In an examination of the inhibition of an enzyme by a drug, I, the following Lineweaver-Burk plot was obtained when the reciprocal of initial velocity ($1/v$) was plotted against the reciprocal of substrate concentration ($1/[S]$) for data obtained in the presence or absence of I.

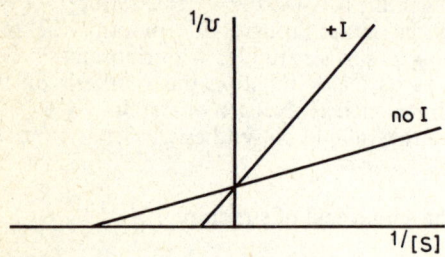

The experimental data obtained in this experiment would be consistent with:

1. irreversible inhibition of the enzyme by I.
2. competitive inhibition of the enzyme by I.
3. the type of inhibition observed when the reaction catalysed by succinate dehydrogenase is performed in the presence of malonate.
4. an increase in the apparent Michaelis constant (K_m) for the substrate.
5. negligible inhibition at very high substrate concentrations relative to the inhibitor.

5.5 Glycogen:

1. is a macromolecule with a variable molecular weight.
2. has a highly branched structure.
3. is a reducing polysaccharide.
4. is the major source of stored energy in the body.
5. is deposited in excessively large amounts in the case of certain inherited metabolic diseases.

6.9 Phosphofructokinase (PFK), catalysing the formation of fructose 1,6-bisphosphate from fructose 6-phosphate:

1. readily catalyses the reverse reaction under physiological conditions.
2. is allosterically activated by ATP.
3. is allosterically inhibited by citrate.
4. is allosterically activated by AMP.
5. is absent from skeletal muscle.

6.10 Uridine diphosphate glucose (UDPG):

1. is a coenzyme derived from a B vitamin.
2. is a substrate for the enzyme glycogen synthase.
3. is an intermediate in the synthesis of starch and cellulose in plants.
4. is required for the normal metabolism of galactose.
5. can be formed directly from UDP-galactose.

7.6 The phosphorylation of ADP to ATP by mitochondria oxidizing succinate:

1. can take place only if the mitochondrial membranes are intact.
2. is driven by chemical coupling to the alternate oxidation and reduction of cytochromes via 'high-energy' intermediates.
3. obtains the necessary energy from the establishment of a proton gradient.
4. is inhibited by valinomycin.
5. is inhibited by rotenone.

8.2 Which of the following statements is/are true of triglycerides (triacyl glycerols)?

1. One gram of anhydrous fat stores one-sixth as much energy as a gram of hydrated glycogen.
2. Hydrolysis in the presence of NaOH will yield 3mol of the sodium salt of the fatty acids ('soaps') and 1mol glycerol.
3. Hydrolysis by pancreatic lipase will yield 3mol of fatty acid and 1mol glycerol.
4. Hydrolysis by the enzyme phospholipase C will yield 3mol of fatty acid and 1mol glycerol.
5. The triacyl glycerol occurring in the cells of adipose tissue represents a mechanical and thermal insulator as well as a store of energy.

10.3 Lipolysis in adipose tissue:

1. is activated by adrenaline.
2. is activated by prostaglandin E1.
3. requires activation of lipoprotein lipase.
4. results in the release of glycerol into the bloodstream.
5. is inhibited by insulin.

12.6 Native DNA:

1. consists of two helical polynucleotide chains coiled round the same axis.
2. consists of two complementary polynucleotide strands.
3. has the sugar-phosphate repeat sequences on the inside of the structure forming the core of the strand.
4. in eukaryotic organisms, is usually found associated with large amounts of basic protein.
5. in prokaryotes, typically consists of one double-stranded circular chromosome.

13.13 During the process of protein biosynthesis in eukaryotic cells:

1. the nucleic acid-containing components required are ribosomes, tRNAs and mRNA.
2. an energy supply in the form of ATP and UTP is required.
3. a supply of NADPH is required.
4. the type of protein synthesized depends only on the type of mRNA present.
5. the absence of even one of the 20 amino acids will prevent the process from proceeding.

14.4 A sphingolipid such as sphingomyelin contains the following:

1. glycerol, phosphate, and two molecules of long chain fatty acid.
2. glycerol, phosphate, two molecules of long chain fatty acid and choline.
3. sphingosine, phosphate, one molecule of long chain fatty acid and choline.
4. sphingosine, phosphate, two molecules of long chain fatty acid and choline.
5. sphingosine, phosphate, two molecules of long chain fatty acid and inositol.

15.1 Membrane lipids:

1. do not contain unsaturated fatty acids.
2. are amphipathic.
3. diffuse easily across the lipid bilayer.
4. may be covalently linked to carbohydrate residues.
5. are arranged symmetrically in the membrane bilayer.

16.1 The metabolism of Ca^{2+} is hormonally regulated by:

1. vitamin D.
2. parathyroid hormone.
3. thyroid hormone.
4. calcitonin.
5. calmodulin.

17.7 Which of the following techniques may be used to quantify a protein antigen in a solution if a specific antiserum is available?

1. Double diffusion in agar (Ouchterlony).
2. Single radial immunodiffusion (Mancini).
3. Haemagglutination-inhibition.
4. Rocket immunoelectrophoresis.
5. The complement fixation test.

18.2 Actin:

1. is an abundant protein in many non-muscle cells.
2. contains a 'hinge' region, like an IgG molecule.
3. is processed by chymotrypsin to form α-actinin.
4. is present in intestinal microvilli.
5. is the major component of coated vesicles.

20.6 Acute hepatitis is normally associated with:

1. a decreased serum GOT/GPT ratio (i.e. glutamate oxaloacetate transaminase/glutamate pyruvate transaminase ratio).
2. a greatly increased serum level of sorbitol dehydrogenase.
3. an increased serum creatine phosphokinase (creatine kinase) activity.
4. an increased serum lactate dehydrogenase activity.
5. an increased serum aldolase activity.

21.1 Carboxylation reactions (CO_2-fixation reactions) in animal cells:

1. require thiamine pyrophosphate.
2. frequently require biotin.
3. usually require energy in the form of a nucleoside triphosphate (e.g. ATP).
4. involve lipoic acid.
5. are often unidirectional and therefore may be important control points in the pathways to which they belong.

22.10 Which of the following features are exhibited by both green plant photosynthesis and bacterial photosynthesis?

1. The presence of two photosystems.
2. The utilization of H_2O as electron donor.
3. The possession of chloroplasts.
4. The utilization of some form of chlorophyll as a light receptor.
5. The production of ATP, utilizing the energy stored in a gradient of protons.

Question Number		Answers			
1.5	1.F	2.T	3.T	4.F	5.T
2.1	1.F	2.T	3.F	4.T	5.F
3.3	1.T	2.T	3.F	4.T	5.F
4.1	1.F	2.T	3.T	4.T	5.T
5.5	1.T	2.T	3.F	4.F	5.T
6.9	1.F	2.F	3.T	4.T	5.F
6.10	1.F	2.T	3.F	4.T	5.T
7.6	1.T	2.F	3.T	4.T	5.F
8.2	1.F	2.T	3.T	4.F	5.T
10.3	1.T	2.F	3.F	4.T	5.T
12.6	1.T	2.T	3.F	4.T	5.T
13.13	1.T	2.F	3.F	4.T	5.T
14.4	1.F	2.F	3.T	4.F	5.F
15.1	1.F	2.T	3.F	4.T	5.F
16.1	1.T	2.T	3.F	4.T	5.F
17.7	1.F	2.T	3.T	4.T	5.T
18.2	1.T	2.F	3.F	4.T	5.F
20.6	1.T	2.T	3.F	4.T	5.T
21.1	1.F	2.T	3.T	4.F	5.T
22.10	1.F	2.F	3.F	4.T	5.T

1.1 Lysosomes:

1. contain their own DNA.
2. contain a store of glycogen.
3. are bounded by a double membrane system.
4. can fuse with endocytotic vesicles.
5. have an acidic internal pH.

3.2 The following pattern was obtained following paper electrophoresis at pH 8.6 with a newly discovered haemoglobin variant, Hb$_{Toytown}$. Normal human haemoglobin, Hb$_A$, was electrophoresed simultaneously for comparison.

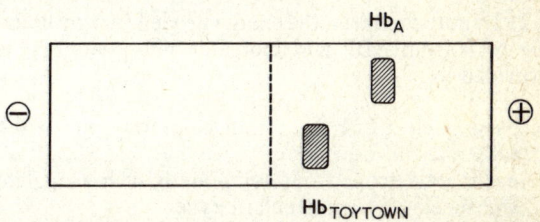

HbₐA

Hb TOYTOWN

From the result of this electrophoresis experiment, it can be concluded that:

1. both haemoglobins have four subunits.
2. a mixture of the two haemoglobins could be separated from one another by ion-exchange chromatography on DEAE-cellulose.
3. the difference in electrophoretic mobility may result from a single amino acid change.
4. it is possible that a negatively charged amino acid residue in Hb$_A$ has been replaced by a neutral one in Hb$_{Toytown}$.
5. it is possible that an uncharged amino acid residue in Hb$_A$ has been replaced by a negatively charged one in Hb$_{Toytown}$.

5.7 The common monosaccharides:

1. contain asymmetric centres.
2. are of two types, aldoses and ketoses.
3. tend to exist as ring structures in solution.
4. include glucose, galactose and raffinose.
5. are all readily formed from glucose by living cells.

6.6   Disaccharides in the diet, or those formed from polysaccharide digestion:

1. are absorbed as such by the small intestine.
2. are hydrolysed to constituent monosaccharides by enzymes present in pancreatic juice.
3. are hydrolysed by specific enzymes of the small intestine.
4. are quantitatively unimportant since there are no enzyme deficiency diseases associated with their metabolism.
5. include lactose which, if not hydrolysed by β-galactosidase (lactase), causes galactosaemia.

7.2   In mitochondria, the energy needed for synthesizing ATP from ADP and inorganic phosphate (P_i) is provided by:

1. passing electrons from molecular oxygen to the substrates of respiration.
2. passing electrons from compounds such as malate and succinate to molecular oxygen.
3. dissipating a concentration gradient of protons across the inner mitochondrial membrane.
4. dissipating a concentration gradient of pyruvate across the inner mitochondrial membrane.
5. converting glucose into lactate.

8.6   3-Hydroxy-3-methylglutaryl CoA (HMG-CoA):

1. is an example of one of the 'ketone bodies'.
2. is formed by the successive condensation of three molecules of acetyl CoA.
3. is an intermediate in the biosynthesis of cholesterol.
4. is an intermediate in the biosynthesis of palmitoyl CoA.
5. is formed in the degradation of leucine.

8.8 The fatty acid synthase complex of mammals contains:

1. covalently bound acyl carrier protein (ACP).
2. covalently bound pyridoxal phosphate.
3. seven pairs of identical subunits.
4. two identical subunits, each having several enzyme activities.
5. two different kinds of subunit, each having several enzyme activities.

9.3 Amino acids are metabolic precursors of a variety of important biomolecules. Mammals can convert:

1. tyrosine into thyroxine.
2. tryptophan into adrenaline.
3. histidine into histamine.
4. tryptophan into the nicotinamide ring of NAD^+.
5. serine into sphingosine.

10.4 Citrate:

1. activates pyruvate carboxylase.
2. activates acetyl CoA carboxylase.
3. activates phosphofructokinase.
4. inhibits pyruvate dehydrogenase.
5. inhibits α-oxoglutarate dehydrogenase.

11.1 The methyl group of N^5-methyltetrahydrofolate ($CH_3 FH_4$) can be directly donated in the:

1. conversion of homocysteine to methionine.
2. biosynthesis of phosphatidyl choline.
3. conversion of glycine to serine.
4. methylation of deoxyuridylate to deoxythymidylate.
5. biosynthesis of the purine ring system.

13.4 The genome of a virus:

1. is known as a plasmid.
2. is always DNA.
3. may be integrated into the 'host' chromosome.
4. may comprise many (>100) genes.
5. may have overlapping genes.

13.11 The biosynthesis of RNA in eukaryotes is inhibited by:

1. actinomycin D.
2. rifamycin.
3. penicillin.
4. α-amanitin.
5. streptomycin.

14.1 Which of the following is true of chylomicrons?

1. The range of diameters encountered is 30–500nm.
2. Chylomicrons contain about 50% lipid and 50% protein.
3. Over 80% of the lipid content of chylomicrons constitutes triacylglycerols.
4. They have a density in the range 1.006–1.063 g/cm^3.
5. They contain small amounts of free cholesterol.

15.3 Integral membrane proteins:

1. can extend all the way through the lipid bilayer.
2. have a relatively high content of hydrophobic residues.
3. always have lipid covalently attached.
4. are called 'lectins' when they contain covalently bound carbohydrates.
5. do not contain regions of α-helix.

16.3 Steroid hormone action:

1. normally involves metabolism of the steroid hormone to an active derivative in the target tissue.
2. involves the second messenger, cyclic AMP.
3. involves selective gene transcription.
4. requires plasma membrane receptors in target cells.
5. involves covalent attachment of the steroid to a receptor.

17.4 Antibody diversity, which enables the body to respond to many different types of antigenic challenge, comes about because:

1. we have one gene for each of the types of antibody molecule we can make.
2. we have a single gene which is susceptible to mutation, producing immunoglobulins with new amino acid sequences, some of which have a complementary shape to the antigen molecule.
3. we have several genes for the different parts of the 'variable' region of the antibody molecule and, by joining these in various ways, extensive diversity is generated.

4. the IgG molecule is flexible because its -Pro-Pro-Pro- sequence in the hinge regions can adapt its shape to fit an antigen molecule.
5. any IgG molecule needs to attach only to a small region of the antigen molecule, and therefore a high degree of specificity is not required.

18.8 The sarcoplasmic reticulum:

1. is the site of protein synthesis in muscle cells.
2. is present in skeletal and cardiac muscle but not in smooth muscle.
3. initiates the contractile process in muscle by binding Ca^{2+} ions.
4. is the site of ATP production in muscle cells.
5. contains a Ca^{2+}-binding protein, calsequestrin.

19.9 Ethanol:

1. is absorbed mainly by the stomach.
2. can readily be converted to glycogen by the body.
3. may provide over half the calorie intake of heavy drinkers and alcoholics.
4. is oxidized to acetaldehyde as the first step in its metabolism.
5. when taken in excess, inhibits the absorption of thiamine.

20.1 An inborn error of metabolism is known in which the enzyme glucose 6-phosphatase is deficient. In patients with this disease, under fasting conditions, the enzyme deficiency would be consistent with:

1. failure of adrenaline to raise blood glucose levels.
2. elevated concentrations of free fatty acids in the plasma.
3. metabolic acidosis.
4. excessive glycogen deposition in the liver.
5. elevated levels of plasma insulin.

22.6 Chloroplasts generate:

1. NADH in light.
2. NADPH in light.
3. $FADH_2$ in light.
4. ATP in light.
5. ATP in darkness if they have first been exposed to a buffer of pH 4 for several hours.

Question Number		Answers			
1.1	1.F	2.F	3.F	4.T	5.T
3.2	1.F	2.T	3.T	4.T	5.F
5.7	1.T	2.T	3.T	4.F	5.T
6.6	1.F	2.F	3.T	4.F	5.F
7.2	1.F	2.T	3.T	4.F	5.F
8.6	1.F	2.T	3.T	4.F	5.T
8.8	1.T	2.F	3.F	4.F	5.T
9.3	1.T	2.F	3.T	4.T	5.T
10.4	1.F	2.T	3.F	4.F	5.F
11.1	1.T	2.F	3.F	4.F	5.F
13.4	1.F	2.F	3.T	4.T	5.T
13.11	1.T	2.F	3.F	4.F	5.F
14.1	1.T	2.F	3.T	4.F	5.T
15.3	1.T	2.T	3.F	4.F	5.F
16.3	1.F	2.F	3.T	4.F	5.F
17.4	1.F	2.F	3.T	4.F	5.F
18.8	1.F	2.F	3.F	4.F	5.T
19.9	1.F	2.F	3.T	4.T	5.F
20.1	1.T	2.T	3.T	4.T	5.F
22.6	1.F	2.T	3.F	4.T	5.T

1.8 Which of the following statements is true of the cell nucleus?

1. The nucleus is bounded by a double membrane.
2. The nuclear membrane is freely permeable to small molecules.
3. The nucleus contains ribosomes that synthesize histones and some other chromatin proteins.
4. The nucleolus is the site of synthesis of messenger RNA (mRNA).
5. Prokaryotic cells have their DNA packaged in 'nucleosomes'.

3.5 The structural protein collagen is found in tendon, bone and cartilage. In comparison with typical globular proteins, it has a number of 'unusual' features. These include:

1. a triple-helical structure.
2. the presence of the amino acid hydroxyproline.
3. the presence of a repeating unit -(Gly-X-Y)- for the majority of its length.
4. an extremely low glycine content.
5. the presence of covalent cross-links between amino acid residues other than cysteine.

4.4 Alcohol dehydrogenase catalyses the general reaction:

$$RCH_2OH + NAD^+ \rightleftharpoons RCHO + NADH + H^+$$

in which RCH_2OH can be almost any alcohol. The Michaelis constant (K_m) of the enzyme for alcohol substrates:

1. is independent of the alcohol substrate oxidized.
2. is independent of enzyme concentration.
3. has the units of substrate concentration.
4. is equal to the affinity of the enzyme for its substrate.
5. would be unchanged in the presence of a non-competitive inhibitor.

5.9 Cellulose:

1. is the main structural component of cell walls of both green plants and bacteria.
2. is a linear polysaccharide consisting of 10 000 or more glucosyl units.

3. is in many respects similar to chitin, the structural material of the insect exoskeleton.
4. is a ready source of energy for all mammals.
5. is the most abundant naturally occurring macro-molecule.

6.5 The absorption of glucose in the digestive tract:

1. occurs in the small intestine.
2. is an energy-requiring process.
3. is stimulated by the hormone glucagon.
4. occurs more rapidly than the absorption of any other sugar.
5. is impaired in cases of diabetes mellitus.

7.8 Creatine kinase (creatine phosphokinase; CPK):

1. functions in liver to degrade phosphocreatine (creatine phosphate).
2. catalyses the hydrolysis of phosphocreatine to creatine and inorganic phosphate.
3. produces ATP from phosphocreatine in muscle.
4. is necessary for the transport of creatine into muscle cells.
5. is released into the blood in large amounts, following muscle damage.

8.5 During β-oxidation of long chain fatty acyl CoA derivatives, energy is released. Which of the following best describe the way in which the energy is released and made available?

1. Both NADH and $FADH_2$ are produced which yield ATP via the electron transport chain.
2. NADPH is produced which yields ATP via the electron transport chain.
3. ATP is produced by substrate-level phosphorylation.
4. NADH and $FADH_2$ result from the entry of acetyl CoA into the TCA cycle and yield ATP via the electron transport chain.
5. Reversal of the reaction:
   fatty acid + CoASH + ATP →
   fatty acyl-CoA + AMP + PP_i
   leads to the production of ATP.

10.1 The activity of the enzyme adenylate cyclase can be potentiated by:

1. cholera toxin.
2. diphtheria toxin.
3. glucagon.
4. cortisol.
5. caffeine.

11.2 With respect to the urea cycle and pyrimidine biosynthesis:

1. carbamoyl phosphate is the source of one nitrogen atom and of the CO_2 utilized in urea synthesis.
2. carbamoyl phosphate and aspartate are the precursors required for the assembly of the pyrimidine ring.
3. the formation of carbamoyl phosphate used for pyrimidine biosynthesis occurs in the cytosol.
4. the nitrogen atom of carbamoyl phosphate used for pyrimidine biosynthesis is derived from glutamine.
5. the enzymes of the urea cycle are all located in mitochondria.

12.7 Foreign DNA may enter a bacterial cell by the processes termed:

1. transcription.
2. transformation.
3. semi-conservative replication.
4. conjugation.
5. transduction.

13.5 Several amino acid residues found in proteins are formed as a result of post-translational events. Such amino acids include:

1. phosphoserine.
2. hydroxylysine.
3. γ-aminobutyrate.
4. glutamine.
5. γ-carboxyglutamate.

13.9 Both prokaryotic and eukaryotic types of mRNA:

1. are extensively modified (edited) before translation can occur.

2. are translated in the $5' \rightarrow 3'$ direction.
3. contain a sequence of a hundred or so adenine (A) residues at the $3'$ end.
4. are associated with a single ribosome at a time during translation.
5. can be polycistronic, i.e. can act as a template for the synthesis of more than one polypeptide chain.

14.7 The uptake of low-density lipoproteins (LDL) by animal cells:

1. involves interaction with specific, cell-surface receptors.
2. occurs by endocytosis.
3. occurs by facilitated diffusion.
4. involves the action of lipoprotein lipase.
5. provides the primary source of cholesterol for these cells.

16.7 Vasopressin:

1. is structurally similar to oxytocin.
2. is synthesized in the hypothalamus.
3. is stored and released from the adenohypophysis (anterior pituitary).
4. regulates water re-absorption by the kidney.
5. may be absent or defective in diabetes mellitus.

17.9 The IgM molecule:

1. has a sedimentation coefficient of 7S.
2. is made up of 10 units, each of which is rather similar to a single IgG molecule.
3. is 'decavalent' in its antigen-combining capacity, i.e. is potentially capable of binding ten antigen molecules.
4. is found in secretions such as tears and saliva, as well as in serum.
5. can be dissociated into its constituent units by treatment with a sulphydryl reagent such as 2-mercaptoethanol.

18.6 Monoamine oxidase:

1. is a flavoprotein.
2. is present in the cytosol.
3. catalyses the oxidation of noradrenaline.

4. catalyses the oxidation of γ-aminobutyrate.
5. is inhibited by the benzodiazepine tranquillizers (e.g. Valium).

20.2 Abnormal storage of glycogen results from deficiencies in:

1. phosphorylase.
2. lactate dehydrogenase.
3. phosphofructokinase.
4. amylo-1,6-glucosidase.
5. glycogen synthase.

20.9 The human haemoglobinopathy, thalassaemia:

1. is another name for sickle-cell disease.
2. results from over-production of normal haemo-globin (HbA) leading to haemolytic anaemia.
3. results from partial or complete failure to produce one or other of the polypeptide chains of haemo-globin leading to haemolytic anaemia.
4. can result from altered haemoglobin metabolism following an attack of malaria.
5. results from a deficiency of one of the enzymes involved in the biosynthesis of the haem moiety of haemoglobin.

21.4 The coenzyme pyridoxal phosphate:

1. is a derivative of niacin (nicotinic acid).
2. can be synthesized from simple carbon, nitrogen and phosphate compounds by most bacteria.
3. is a coenzyme for phosphorylation-dephosphoryl-ation reactions.
4. is a coenzyme for transaminase (aminotransferase) reactions.
5. is a coenzyme for many amino acid decarboxylation reactions.

22.7 The first identifiable product of CO_2 fixation in photosynthesis is:

1. malonyl CoA.
2. ribulose 1,5-bisphosphate.
3. glycerate 3-phosphate (3-phosphoglycerate).
4. glyceraldehyde 3-phosphate.
5. glycerate 1,3-bisphosphate.

Question Number	Answers				
1.8	1.T	2.T	3.F	4.F	5.F
3.5	1.T	2.T	3.T	4.F	5.T
4.4	1.F	2.T	3.F	4.F	5.T
5.9	1.F	2.T	3.F	4.F	5.T
6.5	1.T	2.T	3.F	4.F	5.F
7.8	1.F	2.F	3.F	4.F	5.T
8.5	1.T	2.F	3.F	4.T	5.F
10.1	1.T	2.F	3.T	4.F	5.F
11.2	1.T	2.T	3.T	4.T	5.F
12.7	1.F	2.T	3.F	4.T	5.T
13.5	1.T	2.T	3.F	4.F	5.T
13.9	1.F	2.T	3.F	4.F	5.F
14.7	1.T	2.T	3.F	4.F	5.T
16.7	1.T	2.T	3.F	4.T	5.F
17.9	1.F	2.F	3.T	4.F	5.T
18.6	1.T	2.F	3.T	4.F	5.F
20.2	1.T	2.F	3.F	4.T	5.T
20.9	1.F	2.F	3.T	4.F	5.F
21.4	1.F	2.T	3.F	4.T	5.T
22.7	1.F	2.F	3.T	4.F	5.F

1.7 The Golgi apparatus:

1. is composed of a stack of disc-shaped cisternae.
2. is involved in the biosynthesis of phospholipids.
3. contains the enzymes of the glyoxylate cycle.
4. is involved in the glycosylation of proteins using sugars linked to the lipid, dolichol.
5. packages newly synthesized glycoproteins and directs their transport to intracellular and extracellular destinations.

2.8 The peptide of structure:

1. contains valine as the N-terminal amino acid residue.
2. contains three peptide bonds.
3. would be expected to show hydrophobic properties.
4. would absorb light in the ultraviolet region at about 280nm.
5. would be cleaved by trypsin.

3.6 Different amino acid side chains (R-groups) give different regions of polypeptides their distinct properties. With regard to protein structure:

1. typical globular proteins tend to have hydrophilic amino acid residues on the 'outside' exposed to the aqueous environment, and hydrophobic residues 'inside'.
2. proteins destined to be completely inserted into membranes would be expected to have hydrophobic amino acids on their surfaces.
3. some membrane proteins would be expected to have regions of hydrophobic and regions of hydrophilic amino acids on their surfaces.
4. the cleft into which the oxygen-binding haem group fits in the haemoglobin molecule provides a totally hydrophilic environment.
5. the cleft that forms the active site of the enzyme, lysozyme, and to which bacterial cell wall polysaccharide binds, provides a hydrophobic environment.

4.8   Serum glutamate-oxaloacetate transaminase (GOT):

1. requires biotin as a cofactor.
2. activity is raised after myocardial infarction.
3. can be determined by adding suitable amounts of glutamate, oxaloacetate, lactate dehydrogenase (LDH) and NADH, and by measuring the decrease in absorbance at 340 nm.
4. activity would not be raised after liver damage.
5. is a source of ATP in erythrocytes.

5.10 Consider the structure:

It is:

1. a sugar phosphate ester.
2. a phosphate anhydride.
3. a reducing sugar.
4. glucose 1-phosphate.
5. a ketohexose phosphate.

6.3 The enzyme α-amylase:

1. hydrolyses glycogen to give glucose and maltose.
2. is present in saliva.
3. is activated by trypsin in the small intestine.
4. hydrolyses cellulose to glucose and maltose.
5. if absent from the pancreatic secretion, results in incomplete digestion of starch, a large bacterial population in the faeces and diarrhoea.

7.7 The uptake of oxygen by isolated, coupled mito-chondria:

1. is dependent upon the presence of an oxidizable substrate such as succinate or malate.
2. is greatly stimulated by the addition of ADP.
3. is inhibited by puromycin.
4. is inhibited by 2,4-dinitrophenol.
5. with most substrates, results in the formation of six molecules of ATP for every molecule of oxygen consumed.

8.10 Which of the following compounds is involved in the transport of long chain fatty acids into the mitochondrion across the inner mitochondrial membrane?

1. Oxaloacetate
2. Citrate
3. Carnitine
4. Taurine
5. Acyl carrier protein

9.5 Which of the following compounds participate in or are closely associated with the urea cycle?

1. Ornithine
2. Citrulline
3. Isocitrate
4. Arginino-succinate
5. Phosphatidyl choline

11.7 Mammalian ribonucleotide reductase:

1. catalyses the reduction of a ribonucleoside diphosphate (e.g. ADP) to the corresponding deoxyribonucleoside diphosphate (e.g. dADP).
2. uses NADPH as cofactor.
3. uses vitamin B_{12} as cofactor.
4. uses a protein cofactor such as thioredoxin.
5. is an allosteric enzyme.

12.5 Messenger RNA (mRNA):

1. is about 100 nucleotides long.
2. in bacteria is often long enough to code for more than one polypeptide chain.
3. will always have an AUG sequence near the 5' end.
4. is metabolically stable.
5. when transcribed from viral genes may code for more than one protein from the same nucleotide sequence.

13.14 Protein biosynthesis in prokaryotic cells is inhibited by:

1. tetracycline.
2. cycloheximide.
3. methotrexate.
4. chloramphenicol.
5. theophylline.

14.8 Gangliosides:

1. are phospholipids.
2. contain sialic acid (e.g. *N*-acetylneuraminic acid).
3. are typified by sphingomyelin.
4. are present in plasma membranes.
5. are elevated in Tay-Sachs disease.

15.5 The uptake of glucose by the erythrocyte:

1. is an example of facilitated diffusion.
2. requires hydrolysis of ATP.
3. is saturable.
4. is inhibited by ouabain.
5. is analogous to the uptake of glucose by *Escherichia coli*.

16.6 Thyroxine:

1. is derived by proteolysis from a precursor protein, thyroglobulin.
2. contains iodine.
3. exerts its hormonal effects by increasing intracellular calcium levels.
4. binds specifically to a cell-surface receptor.
5. selectively affects transcription in target tissues.

17.6 Completely monospecific antibody, i.e. one consisting of a single type of molecule, may be produced by:

1. a single lymph node in culture.
2. a myeloma.
3. a hybridoma.
4. immunizing with a single, pure antigen.
5. using immunoaffinity chromatography to isolate the antibody.

18.9 Smooth muscle:

1. is striated in appearance.
2. contains the troponin system.
3. contains tropomyosin.
4. uses Ca^{2+} ions to regulate the contractile process.
5. relaxes in response to an increase in intracellular cyclic AMP.

19.7 Vitamin D, cholecalciferol:

1. can produce toxic effects if taken in excess.
2. is necessary for the absorption of calcium.
3. activates enzymes necessary for the deposition of hydroxylapatite in bones and teeth.
4. under certain circumstances, can be synthesized from cholesterol.
5. absorption from the small intestine is promoted by bile salts.

21.2 Penicillin G, from *Penicillium chrysogenum:*

1. is a polypeptide antibiotic.
2. specifically inhibits protein synthesis by bacterial (70S) ribosomes.
3. specifically inhibits the transpeptidase reaction in bacterial cell-wall synthesis.
4. is effective as an antibiotic only against cells that are actively dividing.
5. is destroyed by penicillinase which confers resistance on a range of bacteria.

22.1 The process of nitrogen fixation in living organisms:

1. occurs only in green plants.
2. occurs only in chlorophyll-containing organisms.
3. is carried out by an iron-molybdenum enzyme complex called 'nitrogenase'.
4. requires the expenditure of 1 mol of ATP for every mol of nitrogen (N_2) fixed as ammonia.
5. requires 'reducing power' (six electrons) to reduce 1 mol of nitrogen to ammonia.

Question Number	Answers				
1.7	1.T	2.F	3.F	4.F	5.T
2.8	1.F	2.F	3.T	4.T	5.F
3.6	1.T	2.T	3.T	4.F	5.F
4.8	1.F	2.T	3.F	4.F	5.F
5.10	1.T	2.F	3.T	4.F	5.F
6.3	1.T	2.T	3.F	4.F	5.T
7.7	1.T	2.T	3.F	4.F	5.T
8.10	1.F	2.F	3.T	4.F	5.F
9.5	1.T	2.T	3.F	4.T	5.F
11.7	1.T	2.T	3.F	4.T	5.T
12.5	1.F	2.T	3.T	4.F	5.T
13.14	1.T	2.F	3.F	4.T	5.F
14.8	1.F	2.T	3.F	4.T	5.T
15.5	1.T	2.F	3.T	4.F	5.F
16.6	1.T	2.T	3.F	4.F	5.T
17.6	1.F	2.T	3.T	4.F	5.F
18.9	1.F	2.F	3.T	4.T	5.T
19.7	1.T	2.T	3.F	4.T	5.T
21.2	1.F	2.F	3.T	4.T	5.T
22.1	1.F	2.F	3.T	4.F	5.T

1.6 When subcellular fractionation of a tissue is carried out, the efficiency of the fractionation process is often monitored by measuring the activities of 'marker enzymes' characteristic of a particular cell compartment, or organelle. Appropriate marker enzymes might include:

1. acid phosphatase for lysosomes.
2. pyruvate dehydrogenase for the cytosol.
3. acetyl CoA carboxylase for the mitochondrial matrix space.
4. glucose 6-phosphatase for endoplasmic reticulum ('microsomes').
5. catalase for peroxisomes.

2.5 The following are sulphur-containing amino acids typically found in proteins:

$$
\begin{array}{cc}
SH & CH_2S-CH_3 \\
| & | \\
CH_2 & CH_2 \\
| & | \\
H_2N-CH-CO_2H & H_2N-CH-CO_2H \\
(A) & (B)
\end{array}
$$

Which of the following statements apply?

1. Amino acid A could form cross-links in proteins.
2. Amino acid B could form cross-links in proteins.
3. The keratin of feather, hair and nail contains relatively high amounts of amino acid A.
4. Amino acid B is an important donor of methyl groups in metabolism.
5. An enzyme that has amino acid A in its active site will be inhibited by treatment with sodium iodo-acetate, ICH_2CO_2Na.

4.3 It is often convenient to assay an enzyme reaction by coupling it with another enzyme to form a product that is easily quantified. An example is the assay of pyruvate kinase by coupling it with lactate dehydrogenase (LDH):

phosphoenolpyruvate + ADP + H^+ → pyruvate + ATP

pyruvate + NADH + H^+ → lactate + NAD^+

In relation to this assay procedure:

1. 1 mol NAD^+ is produced for each mol phospho-enolpyruvate metabolized.

2. of the two enzymes, pyruvate kinase should be in excess.
3. the reaction could be followed by measuring the decrease in absorbance at 340 nm.
4. the reaction could be followed by measuring the decrease in absorbance at 260 nm.
5. the total change in NADH concentration provides a measure of pyruvate kinase activity.

5.8 Polysaccharides:

1. contain many monosaccharide units which may or may not be of the same kind.
2. function mainly as storage or structural compounds.
3. may be easily separated from each other by ion-exchange chromatography or electrophoresis.
4. are present in large amounts in connective tissue.
5. may have amino acid residues attached after completion of the carbohydrate backbone, giving rise to glycoproteins and proteoglycans.

6.11 Mammalian glucose 6-phosphate dehydrogenase:

1. catalyses the oxidative conversion of glucose 6-phosphate to ribulose 5-phosphate.
2. is sometimes congenitally absent in humans, in which case there is a deposition of abnormal amounts of glycogen in the liver.
3. is highly specific for NAD^+ as the hydrogen acceptor.
4. is important in erythrocytes in preventing met-haemoglobinaemia.
5. if absent from erythrocytes, results in primaquine sensitivity.

7.3 The cytochrome P_{450} system:

1. is involved in the hydroxylation of steroids.
2. is involved in the hydroxylation of phenylalanine to tyrosine.
3. is involved in the hydroxylation of prolyl residues in collagen.
4. requires NADH for activity.
5. is located in mitochondria in the adrenal cortex.

8.12 'Ketone bodies' (acetoacetate, β-hydroxy-butyrate and acetone) are found in the blood of humans in large amounts:

1. immediately following a prolonged bout of vigorous exercise.
2. in untreated phenylketonuria.
3. in uncontrolled diabetes mellitus.
4. during starvation.
5. following a bout of deep breathing.

9.8 Choline:

1. is a quaternary amine.
2. can function as a source of methyl groups.
3. is a precursor of cholic acid.
4. is present, as phosphatidyl choline, in mitochondrial membranes.
5. is the neurotransmitter acting at the mammalian neuromuscular junction.

10.2 Calcium ions, Ca^{2+}:

1. inhibit the action of the regulatory protein, calmodulin.
2. activate phosphorylase kinase.
3. activate the dependent- (D- or b-) form of glycogen synthetase.
4. are required for the blood clotting cascade.
5. are required for the release of noradrenaline from neurons.

10.6 The biosynthesis of many cell constituents such as amino acids, purines and pyrimidines:

1. normally begins with a pathway-specific enzyme that is essentially unidirectional.
2. is usually subject to control by negative-feedback inhibition.
3. in bacteria, is frequently regulated by induction of the necessary enzymes.
4. will decrease when ATP levels are high.
5. will invariably involve a reduction reaction at some stage in the pathway.

11.4 The hydrolysis of ATP to AMP occurs in the overall reaction catalysed by:

1. (Na^+/K^+)-ATPase.
2. acyl CoA synthetase.

3. adenylate cyclase.
4. argininosuccinate synthetase.
5. aminoacyl-tRNA synthetase.

12.4 The extra-chromosomal DNA present in many bacteria:

1. exists in the form of plasmids.
2. is single-stranded.
3. frequently carries genes that specify antibiotic resistance.
4. can often be passed between cells of different species.
5. may confer upon the cell the ability to take part in conjugation.

13.7 Histones:

1. are the major proteins present in chromatin.
2. are acidic proteins.
3. are coded in the DNA in multiple copies.
4. have amino acid sequences that have been highly conserved during evolution.
5. constitute the nuclear receptors for steroid hormones.

14.5 The overall pathway of prostaglandin biosynthesis may be represented as follows:

$$\text{Membrane phospholipids} \xrightarrow{\text{(A)}} \text{arachidonic acid} \xrightarrow{\text{(B)}} \text{prostaglandins}$$

Which of the following is true of this pathway?

1. The enzyme involved at step A is phospholipase A.
2. The reaction, A, is stimulated by anti-inflammatory steroids.
3. The enzyme involved at step B is cyclo-oxygenase.
4. The reaction, B, is stimulated by aspirin.
5. The reaction, A, is the rate-limiting step in prostaglandin synthesis.

16.8 A deficiency of thyroid hormone may result from:

1. phaeochromocytoma.
2. Cushing's syndrome.
3. Lesch-Nyhan syndrome.
4. Graves' disease.
5. Hashimoto's disease.

18.5 Compounds which function as neurotransmitters include:

1. enkephalin.
2. carnitine.
3. cysteine.
4. γ-aminobutyrate (GABA).
5. 3,4-dihydroxyphenylalanine (DOPA).

19.8 A lack of functional 'intrinsic factor', a glycoprotein secreted by the stomach:

1. is caused by a deficiency of dietary vitamin B_{12}.
2. is the most common cause of vitamin B_{12} deficiency.
3. can be compensated for by the administration of vitamin B_{12}.
4. causes iron-deficiency anaemia.
5. is characterized by the presence of high levels of methylmalonate in blood and urine.

20.7 A myocardial infarction will lead to an increase in serum levels of:

1. lactate dehydrogenase.
2. creatine phosphokinase (creatine kinase).
3. glucose 6-phosphatase.
4. glutamate-oxaloacetate transaminase (GOT).
5. glutamate-pyruvate transaminase (GPT).

21.8 The red cells of blood have a finite lifetime, and the degradation of haem in humans:

1. is incomplete in cases of porphyria leading to the accumulation of uroporphyrinogen I or porphobilinogen.
2. gives rise to bile salts.
3. gives rise to bile pigments.
4. involves the spleen and liver.
5. occurs as a result of the turnover of erythrocytes.

22.4 Although photosynthesis occurs in a number of distinct steps, the overall process can be summed up by a relatively simple equation. Which of the following equations summarizes the processes of photosynthesis either in green plants or bacteria? In the equations, (CH_2O) represents carbohydrate.

1. $CO_2 + H_2O \rightarrow (CH_2O) + O_2$
2. $CO_2 + 2H_2O \rightarrow (CH_2O) + O_2 + H_2O$
3. $CO_2 + H_2S \rightarrow (CH_2O) + S + \frac{1}{2}O_2$
4. $CO_2 + 2H_2S \rightarrow (CH_2O) + 2S + H_2O$
5. $CO_2 + 2\text{ lactate} \rightarrow (CH_2O) + 2\text{ pyruvate} + H_2O$

Question Number		Answers			
1.6	1.T	2.F	3.F	4.T	5.T
2.5	1.T	2.F	3.T	4.T	5.T
4.3	1.T	2.F	3.T	4.F	5.F
5.8	1.T	2.T	3.F	4.T	5.F
6.11	1.F	2.F	3.F	4.T	5.T
7.3	1.T	2.F	3.F	4.F	5.T
8.12	1.T	2.F	3.T	4.T	5.F
9.8	1.T	2.T	3.F	4.T	5.F
10.2	1.F	2.T	3.F	4.T	5.T
10.6	1.T	2.T	3.F	4.F	5.T
11.4	1.F	2.T	3.F	4.T	5.T
12.4	1.T	2.F	3.T	4.T	5.T
13.7	1.T	2.F	3.T	4.T	5.F
14.5	1.T	2.F	3.T	4.F	5.T
16.8	1.F	2.F	3.F	4.T	5.T
18.5	1.T	2.F	3.F	4.T	5.F
19.8	1.F	2.T	3.T	4.F	5.T
20.7	1.T	2.T	3.F	4.T	5.T
21.8	1.F	2.F	3.T	4.T	5.T
22.4	1.F	2.T	3.F	4.T	5.T

1.2 Microtubules:

1. are composed of filaments of actin.
2. do not assemble in the presence of the alkaloid colchicine.
3. form a major component of the mitotic spindle.
4. form a major component of cilia and flagella of eukaryotic cells.
5. have ATPase activity.

2.2 The structures A–E are of five amino acids typically found in proteins.

$$CH_3 \quad CH_3$$
$$\diagdown \diagup$$
$$CH$$
$$CH_3 \qquad \qquad | \qquad \qquad CH_2OH$$
$$| \qquad \qquad | \qquad \qquad |$$
$$H_2N-CH-CO_2H \quad H_2N-CH-CO_2H \quad H_2N-CH-CO_2H$$

(A)   (B)   (C)

$$CH_2S-CH_3 \qquad CH_2-CH_2$$
$$| \qquad \qquad \quad | \qquad \quad |$$
$$CH_2 \qquad \qquad CH_2 \quad CH-CO_2H$$
$$| \qquad \qquad \quad \diagdown \quad \diagup$$
$$H_2N-CH-CO_2H \qquad \quad N$$
$$H$$

(D)   (E)

These amino acids:

1. all have hydrophobic side chain (R-) groups.
2. all have non-ionizable side chain (R-) groups.
3. all can participate in α-helix formation in proteins.
4. are unable to form cross-links in proteins.
5. would be found in the L-form in proteins.

3.4 The following result was obtained in an experiment in which immunoglobulin G (IgG) was subjected to electrophoresis in the presence of sodium dodecyl sulphate (SDS) and 2-mercaptoethanol. A set of standard reference proteins was electrophoresed simultaneously. The reference proteins were A, bovine serum albumin (M_r = 68 000); B, ovalbumin (M_r = 45 000); C, chymotrypsin (M_r = 25 000); D, cytochrome c (M_r = 12 000).

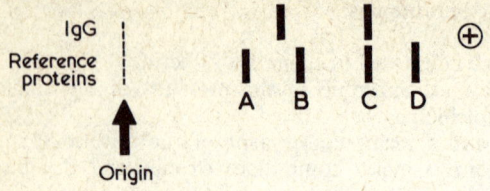

IgG
Reference proteins

Origin

A B C D ⊕

From these results it can be concluded that:

1. the M_r of IgG is 150 000.
2. IgG is made up of polypeptides of approximate M_r 25 000 and 50 000.
3. the M_r of IgG is likely to be 75 000 or some multiple of 75 000.
4. IgG is a glycoprotein.
5. the polypeptides of IgG are joined by disulphide bonds.

4.2 The reaction catalysed by chymotrypsin:

1. involves hydrolysis of a peptide bond on the C-terminal side of arginyl or lysyl residues.
2. takes place at a serine residue in the active site.
3. involves formation of a covalent acyl-enzyme intermediate.
4. has an optimal pH of 1.5–2.5.
5. is inhibited by di-isopropylphosphofluoridate (Dip-F or DFP).

5.6 Examples of important heteropolysaccharides are:

1. amylopectin.
2. heparin.
3. the polysaccharide component of murein (peptido-glycan).
4. keratin.
5. hyaluronic acid.

6.4 The enzyme β-amylase:

1. hydrolyses the β-1,4 glycosidic linkages of cellulose.
2. hydrolyses starch to maltose.
3. is present in large amounts in certain germinating seedlings.
4. is present in saliva.
5. is present in mammalian lysosomes.

8.1 The naturally-occurring compound of structure:

$$CH_3(CH_2)_4(CH=CH.CH_2)_2(CH_2)_6CO_2H$$

1. is linolenic acid.
2. is essential in the diet of humans.
3. would have its double bonds in the *trans* configuration.
4. would have a melting point higher than that of the corresponding fully saturated C_{18} acid.
5. is a precursor of prostaglandins.

8.11 Fatty acid biosynthesis involves:

1. malonic acid.
2. malonyl CoA.
3. acyl carrier protein.
4. coenzyme A.
5. sulphydryl groups.

9.7 The diagram shows a simplified form of the urea cycle.

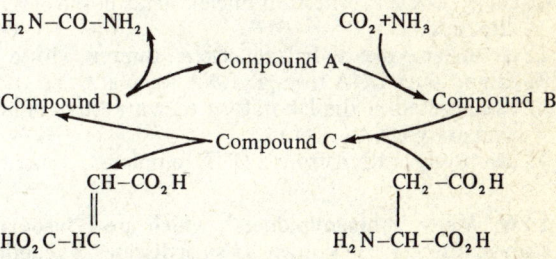

In this cycle:

1. compound A is arginine.
2. compound D is citrulline.
3. compound C is argininosuccinate.
4. the compound formed as a result of the conversion of compound C to D could enter the tricarboxylic acid cycle.
5. energy has to be supplied in the form of ATP.

10.7 The induction of β-galactosidase in *Escherichia coli*:

1. is under the control of the *gal* operon.
2. is brought about only by lactose.
3. is accompanied by the induction of a transport system for lactose.
4. is mediated through cyclic $3'5'$-AMP.
5. ceases if glucose is added to the growth medium.

11.5 Guanosine triphosphate (GTP) is involved in the:

1. activation of adenylate cyclase by glucagon.
2. activation of Na^+ influx by acetylcholine at the neuromuscular junction.
3. elongation stage of protein biosynthesis.
4. reaction catalysed by phosphoenolpyruvate carboxykinase.
5. reaction catalysed by pyruvate carboxylase.

13.2 The 'signal sequence' of secretory proteins:

1. specifies the point of attachment of carbohydrate residues.
2. directs the nascent protein into the endoplasmic reticulum.
3. determines the final conformation of the active protein.
4. is located at the C-terminus of the protein.
5. has a hydrophobic character.

13.3 Reverse transcriptase:

1. catalyses the synthesis of nucleic acids in the $3'{\rightarrow}5'$ direction.
2. is an enzyme coded by DNA tumour viruses.
3. can produce DNA from an RNA template.
4. can be used in the laboratory to synthesize complementary DNA (cDNA).
5. can produce RNA from a DNA template.

14.6 Many 'sphingolipidoses', which are lysosomal storage diseases, are known. In such diseases, a genetic error results in the failure to catabolize a particular lipid or lipid metabolite. An example is Tay-Sachs disease. Which of the following applies to the enzymes involved in sphingolipid catabolic pathways?

1. The enzymes in question are hydrolases.
2. the pH optimum of each of the enzymes is in the range pH 7–10.
3. The enzymes in question are glycoproteins.
4. Usually a collection of sphingolipids accumulates when there is a genetic error of the type mentioned above.
5. Most of the enzymes occur as isoenzymes.

16.4 Angiotensin II:

1. is derived from angiotensin I by proteolysis.
2. produces thirst when administered centrally to an animal.
3. is formed in the kidney.

4. regulates the biosynthesis and secretion of cortisone.
5. regulates the biosynthesis and secretion of aldosterone.

18.3 The naturally occurring peptide, methionine-enkephalin (Tyr-Gly-Gly-Phe-Met):

1. could be hydrolysed by trypsin.
2. could be hydrolysed by cyanogen bromide.
3. is synthesized by a non-ribosomal mechanism.
4. causes the secretion of corticotropin (ACTH) from the pituitary.
5. is the compound normally binding to the membrane receptors that recognize morphine.

19.5 Which of the following diseases may result from a dietary deficiency?

1. Pernicious anaemia
2. Xerophthalmia
3. Marasmus
4. Simple goitre
5. Addison's disease

20.8 An elevated serum alkaline phosphatase level is associated with:

1. Paget's disease.
2. muscular dystrophy.
3. bone tumours.
4. rickets.
5. carcinoma of the prostate.

21.5 Both haemoglobin and cytochrome c:

1. consist of protein, and an iron-containing prosthetic group.
2. carry oxygen.
3. contain iron which is oxidized in each case to the ferric state, Fe(III), during normal physiological function.
4. are characterized by a strong absorption peak close to 410 nm.
5. have similar relative molecular masses (M_r values).

22.5 Which of the following take part in the 'light phase' of photosynthesis?

1. Ferredoxin
2. Cytochrome P_{450}
3. Plastoquinone
4. Plastocyanin
5. Haemocyanin

Question Number		Answers			
1.2	1.F	2.T	3.T	4.T	5.F
2.2	1.F	2.T	3.F	4.T	5.T
3.4	1.F	2.T	3.T	4.F	5.F
4.2	1.F	2.T	3.T	4.F	5.T
5.6	1.F	2.T	3.T	4.F	5.T
6.4	1.F	2.T	3.T	4.F	5.F
8.1	1.F	2.T	3.F	4.F	5.T
8.11	1.F	2.T	3.T	4.T	5.T
9.7	1.F	2.F	3.T	4.T	5.T
10.7	1.F	2.F	3.T	4.T	5.T
11.5	1.T	2.F	3.T	4.T	5.F
13.2	1.F	2.T	3.F	4.F	5.T
13.3	1.F	2.F	3.T	4.T	5.F
14.6	1.T	2.F	3.T	4.F	5.T
16.4	1.T	2.T	3.F	4.F	5.T
18.3	1.F	2.F	3.F	4.F	5.T
19.5	1.F	2.T	3.T	4.T	5.F
20.8	1.T	2.F	3.T	4.T	5.T
21.5	1.T	2.F	3.F	4.T	5.F
22.5	1.T	2.F	3.T	4.T	5.F

54

2.6 Consider the following compound:

$$CO_2H$$
$$|$$
$$S-S-CH_2CH-NH_2 \qquad CH_2CONH_2$$
$$| \qquad\qquad\qquad |$$
$$CH_2 \qquad\qquad CH_3 \qquad\qquad CH_2$$
$$| \qquad\qquad\quad | \qquad\qquad\quad |$$
$$H_2N-CH_2-CO-NH-CH-CO-NH-CH-CO-NH-CH-CO_2H$$

It contains:

1. four peptide bonds.
2. one amide bond in addition to the peptide bonds.
3. two disulphide bonds.
4. five amino acid residues.
5. glycine as its carboxy-terminal amino acid.

3.7 Which of the following structural proteins are extracellular?

1. Keratin
2. Collagen
3. Elastin
4. Fibronectin
5. Tubulin

4.7 Allosteric enzymes, which typically control the rate of throughput in a metabolic pathway:

1. generally show a hyperbolic dependence of initial rate on substrate concentration.
2. are always oligomeric.
3. alter the equilibrium position of the reaction they catalyse.
4. catalyse the final reaction in a metabolic pathway.
5. are typified by aspartate transcarbamylase (ATCase) from *Escherichia coli*.

5.2 For aldohexoses (e.g. glucose), which have the formula $CHO(CHOH)_4CH_2OH$:

1. there are five asymmetric carbon atoms in the straight chain form.
2. there are 16 different molecular straight-chain structures and therefore 16 chemically different sugars.
3. there are 16 different straight-chain structures and therefore 32 different ring structures.
4. D-glucose differs from L-glucose only in the configuration of groups on carbon-5.
5. all D-aldohexose sugars rotate plane-polarized light to the right.

6.7 Glucuronic acid:

1. is formed from glucose by oxidation at carbon-1.
2. is important in the formation of certain detoxification products.
3. is present in a number of polysaccharides.
4. is involved in the excretion of bilirubin.
5. is an intermediate in the biosynthesis of vitamin C.

8.4 The following sequence shows the key reactions in the β-oxidation pathway for the catabolism of fatty acids:

$$-CH_2\ CH_2\ COCoA$$

$$\downarrow (A)$$

$$-CH=CHCOCoA$$

$$\downarrow (B)$$

$$\begin{array}{c} -CH-CH_2\ COCoA \\ | \\ OH \end{array}$$

$$\downarrow (C)$$

$$\begin{array}{c} -C-CH_2\ COCoA \\ \| \\ O \end{array}$$

$$CoASH \searrow (D)$$

$$\begin{array}{c} -C-CoA + CH_3\ COCoA \\ \| \\ O \end{array}$$

In this sequence of reactions:

1. steps A and C are dehydrogenations.
2. steps B and D are hydrolyses.
3. steps B and C are stereospecific.
4. step D requires energy in the form of ATP.
5. steps A–D take place in the cytosol of eukaryotic cells.

9.1 Nitrogen fixation (conversion of N_2 to NH_4^+):

1. occurs in all plants.
2. requires the nitrogenase complex.
3. requires at least 12 ATP per N_2 fixed.
4. requires a powerful reductant.
5. requires the participation of CO_2.

11.8 Thymidylate synthetase:

1. catalyses the methylation of deoxyuridine triphosphate (dUTP) to form deoxythymidine triphosphate (dTTP).
2. uses methyltetrahydrofolate as methyl donor.
3. uses NADPH as cofactor.
4. uses a protein cofactor, thioredoxin.
5. is inhibited by fluorodeoxyuridylate (F-dUMP).

13.8 Mitochondrial DNA:

1. normally exists in a circular double-stranded form.
2. is associated with histones.
3. codes for the rRNA of mitochondrial ribosomes.
4. codes for the mRNA species that specify the enzymes of the citric acid (TCA) cycle.
5. uses exactly the same genetic code as nuclear DNA.

13.12 Bacterial restriction endonucleases:

1. can hydrolyse single-stranded DNA molecules.
2. usually hydrolyse DNA adjacent to 'modified' (methylated) bases.
3. hydrolyse DNA molecules foreign to the host cell.
4. recognize short, specific base sequences within the DNA molecule.
5. are used in experiments involving the construction of recombinant DNA molecules.

14.2 The compound shown below is derived from glycerol.

$$
\begin{array}{l}
\text{CH}_2-\text{OCO}-\text{R}_1 \\
\quad | \\
\text{R}_2-\text{CO.O}-\text{CH} \quad\quad \text{O} \\
\quad\quad\quad\quad | \quad\quad\quad\quad \| \\
\quad\quad\quad\text{CH}_2-\text{O}-\text{P}-\text{O}-\text{R}_3 \\
\quad\quad\quad\quad\quad\quad\quad\quad | \\
\quad\quad\quad\quad\quad\quad\quad\quad \text{O}^-
\end{array}
$$

Which of the following enzymes would be expected to catalyse the cleavage of this molecule?

1. Pancreatic lipase
2. Snake venom phospholipase (phospholipase A_2)
3. Plant phospholipase (phospholipase D)
4. Acid phosphatase
5. Lipophosphodiesterase (phospholipase C)

15.6 The glucose transport system of the epithelial cells of the small intestine:

1. can also transport fructose.
2. is inhibited by 2,4-dinitrophenol.
3. is inhibited by phloridzin.
4. is stimulated by insulin.
5. is dependent on co-transport of Na^+.

16.9 Corticotropin (ACTH):

1. is synthesized in the hypothalamus.
2. is synthesized as a multi-hormone precursor.
3. stimulates formation of cyclic AMP in the adrenal cortex.
4. promotes lipolysis in adipose tissue.
5. stimulates the conversion of cholesterol into pregnenolone.

17.5 In the lymphatic system, antibodies for circulation in the serum are produced by:

1. all macrophages.
2. macrophages that have ingested antigen or fragments of antigen.
3. T lymphocytes in the lymph nodes.
4. B lymphocytes that have been transformed into 'plasma cells'.
5. reticulocytes, in the spleen and bone marrow, before they lose their nuclei and become circulating erythrocytes.

17.8 The complement system of the blood:

1. enables antibody molecules to be transported across the placenta.
2. recognizes antibody-antigen complexes and ensures that they are 'marked' for removal from the circulation.
3. recognizes antibody molecules attached to invading micro-organisms ensuring that such cells will be lysed and destroyed.
4. amplifies the response of antibodies to foreign antigens.
5. recognizes cell-wall material in invading and possibly dangerous organisms (e.g. cell wall polysaccharide of yeasts), setting off a sequence of events that will lead to the invader being inactivated.

18.4 Patients with McArdle's disease (a deficiency of muscle glycogen phosphorylase):

1. accumulate excessive amounts of glycogen in muscle.
2. inherit the disease in an autosomal recessive manner.
3. rarely survive beyond the age of two years.
4. would show little or no rise in blood sugar levels after administration of adrenaline.
5. would show little or no rise in blood lactate levels after vigorous exercise.

19.2 The approximate, recommended daily requirements for vitamins are as follows:

1. thiamine; 1.5mg.
2. niacin (nicotinic acid); 2mg.
3. cobalamin; 3mg.
4. ascorbic acid; 4.5mg.
5. folic acid; 0.4mg.

19.4 Iron is:

1. absorbed in the oxidized, ferric Fe(III) form.
2. absorbed very efficiently from the intestine.
3. transported in the blood bound to albumin.
4. stored in the body in the form of ferritin.
5. excreted in large amounts in bile and faeces.

21.10 In the process of porphyrin biosynthesis:

1. four histidine molecules combine to form a tetrapyrrole.
2. the first identifiable step is the enzyme-catalysed combination of glycine and succinyl CoA to form δ-aminolaevulinic acid.
3. four molecules of δ-aminolaevulinic acid combine to form porphobilinogen.
4. pyridoxal phosphate is required as a cofactor.
5. the same initial steps of the pathway are followed whether the porphyrin is incorporated into haemoglobin, cytochromes, the chlorophylls or vitamin B_{12}.

22.8 The operation of the 'C$_4$ pathway' of photosynthesis in certain tropical plants:

1. allows efficient photosynthesis without the need for ATP.
2. achieves high rates of photosynthesis, even with the stomata closed, preventing excessive water loss.
3. achieves maximum rates of photosynthesis with the stomata open in high light intensity.
4. produces oxaloacetate, which is quantitatively converted into carbohydrate.
5. allows photosynthesis to continue in the hours of darkness.

Question Number					
2.6	1.F	2.T	3.F	4.T	5.F
3.7	1.F	2.T	3.T	4.T	5.F
4.7	1.F	2.T	3.F	4.F	5.T
5.2	1.F	2.F	3.T	4.F	5.F
6.7	1.F	2.T	3.T	4.T	5.T
8.4	1.T	2.F	3.T	4.F	5.F
9.1	1.F	2.T	3.T	4.T	5.F
11.8	1.F	2.F	3.F	4.F	5.T
13.8	1.T	2.F	3.T	4.F	5.F
13.12	1.F	2.F	3.T	4.T	5.T
14.2	1.F	2.T	3.T	4.F	5.T
15.6	1.F	2.T	3.T	4.F	5.T
16.9	1.F	2.T	3.T	4.T	5.T
17.5	1.F	2.F	3.F	4.T	5.F
17.8	1.F	2.T	3.T	4.F	5.T
18.4	1.T	2.T	3.F	4.F	5.T
19.2	1.T	2.F	3.F	4.F	5.T
19.4	1.F	2.F	3.F	4.T	5.F
21.10	1.F	2.T	3.F	4.T	5.T
22.8	1.F	2.T	3.F	4.F	5.F

Part 2: Questions and Answers with Explanations

Section 1

CELL STRUCTURE AND FUNCTION

1.1 Lysosomes:

1. contain their own DNA.
2. contain a store of glycogen.
3. are bounded by a double membrane system.
4. can fuse with endocytotic vesicles.
5. have an acidic internal pH.

1. False
Lysosomes are organelles containing a battery of hydrolytic enzymes (e.g. nucleases, proteases, glycosidases, lipases) for *degrading* a range of macromolecules, including nucleic acids.

2. False
See (1). The lysosomal enzyme α-1,4-D-glucosidase (acid maltase) would rapidly degrade any glycogen in lysosomes. An inherited deficiency of this enzyme (Pompe's disease) leads to the accumulation of glycogen in liver and other tissues. The disease is usually fatal in the first year of life.

3. False
Lysosomes are bounded by a single membrane bilayer. Mitochondria, chloroplasts and the cell nucleus, in contrast, are bounded by a double membrane system.

4. True
Fusion with endocytotic vesicles produces a secondary lysosome, sometimes called a digestive vacuole. The endocytosed material is then digested by the lysosomal hydrolases.

5. True
The lysosomal hydrolases have optimal activity at about pH 5. This pH value is maintained within the lysosomes through the action of a H^+-transporting membrane ATPase. The acid pH optimum of the hydrolases protects the cell cytoplasm from damage should any of the lysosomal enzymes leak out.

1.2 Microtubules:

1. are composed of filaments of actin.
2. do not assemble in the presence of the alkaloid colchicine.
3. form a major component of the mitotic spindle.
4. form a major component of cilia and flagella of eukaryotic cells.
5. have ATPase activity.

1. False
Microtubules are one of several types of filamentous protein material found in eukaryotic cells. They are composed of a helical array of α- and β- tubulin sub-units; α- and β-tubulins are both globular proteins of relative molecular mass (M_r) approx. 55 000.

2. True
The alkaloid colchicine prevents the polymerization of tubulin monomers. It will therefore inhibit cellular activities that require intact microtubules.

3. True
Colchicine will therefore prevent cell division and the cells will be held at the metaphase stage.

4. True
The bundle of fibres (axoneme) comprising cilia and flagella are arranged as nine pairs of microtubules surrounding two individual microtubules ('9 + 2' array). Bacterial flagella, composed of flagellin subunits, do not have this structure.

5. False
The energy for movement of eukaryotic cilia and flagella comes from an associated protein called dynein which has ATPase activity.

1.3 The diagram shows a cross-section of a mitochondrion. Identify the structures labelled with letters.

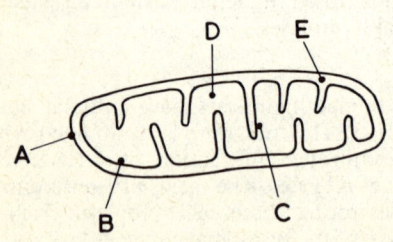

1. Structure A is the cell wall.
2. Structure B is the inner membrane.
3. Structure C is a chloroplast.
4. Structure D is 'matrix'.
5. Structure E is the space where the enzymes of oxidative phosphorylation are found.

1. False
This is the outer membrane, which is smooth and not infolded, and is relatively permeable to metabolites and cofactors.

2. True
This is the inner membrane which is infolded to form cristae (see 3). The inner membrane is impermeable to most substances unless specific transport systems exist.

3. False
This is one of the cristae (see 2). In general, tissues with a high level of metabolic activity have numerous mitochondria in their cells and these mitochondria have many cristae. Mitochondria do not contain chloroplasts and chloroplasts are at least as large as mitochondria, if not larger.

4. True
The gel-like matrix contains many enzymes, cofactors and substrates. The key oxidative reactions occur in the matrix and on the inner membrane.

5. False
This is the intermembrane space and its contents probably differ little from those of the cytoplasmic compartment.

1.4 Mitochondria:

1. are found in all prokaryotic cells, except photosynthetic bacteria.
2. are found in all eukaryotic cells.
3. contain DNA.
4. contain ribosomes.
5. in evolutionary terms, may have originated from primitive bacteria.

1. False
Mitochondria are characteristically absent from prokaryotic cells. A 'typical' mitochondrion is about the same size as a 'typical' bacterium. The functions of mitochondria in prokaryotic cells are carried out by infoldings of the cell membrane.

2. False

Although the presence of mitochondria is one of the distinguishing features of eukaryotes, these organelles are not found in all cells. They are not found in red blood cells nor in certain anaerobic protozoa and fungi that obtain energy exclusively by fermentation.

3. True

Mitochondria contain a small amount of circular DNA which codes for a limited number of mitochondrial proteins.

4. True

In addition to small amounts of DNA, mitochondria also contain protein-synthesizing machinery, including ribosomes. These ribosomes (70S) are similar in size to bacterial ribosomes and smaller than the eukaryotic cytosolic ribosomes (80S).

5. True

The presence of circular DNA, bacterial-type (70S) ribosomes, and their size, all point to mitochondria having evolved from free-living bacteria which at some stage established a symbiotic relationship with a primitive eukaryotic cell.

1.5 The endoplasmic reticulum:

1. forms many folds and convolutions within the space inside mitochondria.
2. contains spaces, or cisternae, which serve as channels for the transport of various products through the cell, usually to its exterior.
3. may be either 'rough' or 'smooth', the former having its surface studded with ribosomes.
4. has as its major function the synthesis of carbohydrates such as glycogen or starch.
5. has a role in lipid biosynthesis.

1. False

It does have many folds and convolutions but it is found within the cytoplasmic space.

2. True

Secreted proteins are synthesized by ribosomes attached to the endoplasmic reticulum (known as the rough endoplasmic reticulum). These proteins pass through the membrane and eventually find their way to the exterior of the cell after being packaged in the Golgi apparatus.

3. True
Smooth endoplasmic reticulum has no attached ribosomes (see 2).

4. False
The endoplasmic reticulum is mainly concerned with the biosynthesis and export of proteins, but see 5.

5. True
The endoplasmic reticulum does have a role in lipid synthesis, particularly in the case of membrane phospholipids.

1.6 When subcellular fractionation of a tissue is carried out, the efficiency of the fractionation process is often monitored by measuring the activities of 'marker enzymes' characteristic of a particular cell compartment, or organelle. Appropriate marker enzymes might include:

1. acid phosphatase for lysosomes.
2. pyruvate dehydrogenase for the cytosol.
3. acetyl CoA carboxylase for the mitochondrial matrix space.
4. glucose 6-phosphatase for endoplasmic reticulum ('microsomes').
5. catalase for peroxisomes.

1. True
Lysosomes contain a battery of hydrolytic enzymes including acid phosphatase.

2. False
Pyruvate dehydrogenase, and hence the formation of acetyl CoA, is located within mitochondria.

3. False
Acetyl CoA carboxylase and fatty acid synthesis occur in the cytosol. For convenience of assay, lactate dehydrogenase is usually chosen as a marker enzyme for the cytosol.

4. True
Glucose 6-phosphatase is associated with the endoplasmic reticulum but is only present in some tissues (e.g. liver, kidney, intestine).

5. True
In addition to catalase, peroxisomes contain a number

of oxidative enzymes producing hydrogen peroxide (e.g. D-amino acid oxidase). Catalase reduces H_2O_2 to water by removal of oxygen and therefore prevents the accumulation of a powerful oxidizing agent (H_2O_2) within the cell.

1.7 The Golgi apparatus:

1. is composed of a stack of disc-shaped cisternae.
2. is involved in the biosynthesis of phospholipids.
3. contains the enzymes of the glyoxylate cycle.
4. is involved in the glycosylation of proteins using sugars linked to the lipid, dolichol.
5. packages newly synthesized glycoproteins and directs their transport to intracellular and extra-cellular destinations.

1. True
The Golgi stacks are usually located near to the cell nucleus and are seen in close association with many specialized, membrane-bound vesicles ('coated vesicles') which transport material to and from the Golgi.

2. False
The biosynthesis of phospholipids is a principal function of the endoplasmic reticulum, not the Golgi apparatus.

3. False
The glyoxylate cycle comprises a specialized set of reactions for converting acetyl CoA to succinate. These reactions occur in germinating seeds allowing conversion of stored seed lipids to sugars. They are also present in bacteria but not in animal cells, which are unable to convert fats into sugars.

4. False
Glycosylation of proteins occurs at two separate intracellular sites. 'Core' sugars are added to newly synthesized proteins in the endoplasmic reticulum. It is this process that uses dolichol-linked sugars as intermediates. Further processing and glycosylation then occurs in the Golgi but, at this site, nucleotide-linked sugars are used as 'activated' intermediates.

5. True
A key function of the Golgi apparatus is directing the intracellular transport of newly synthesized glycoproteins. The precise mechanism is unclear,

but it is probably the nature of the associated sugars that distinguishes proteins destined for export from those directed to lysosomes, mitochondria, etc.

1.8 Which of the following statements is true of the cell nucleus?

1. The nucleus is bounded by a double membrane.
2. The nuclear membrane is freely permeable to small molecules.
3. The nucleus contains ribosomes that synthesize histones and some other chromatin proteins.
4. The nucleolus is the site of synthesis of messenger RNA (mRNA).
5. Prokaryotic cells have their DNA packaged in 'nucleosomes'.

1. True
Chromatin and other contents of the nucleus are separated from the cytoplasm by a double membrane system called the 'nuclear envelope'. The outer nuclear membrane appears to be continuous with the endoplasmic reticulum.

2. True
The eukaryotic nuclear membrane is perforated by 'pores' which allow ready passage of small molecules, including small proteins, between the nucleus and the cytoplasm.

3. False
Although ribosomal subunits are assembled in the nucleus, there are no functional ribosomes within, and no protein synthesis occurs. Histones, and all other nuclear proteins, are imported into the nucleus from the cytoplasm. Special transport mechanisms must exist to allow the entry of large proteins such as DNA polymerase.

4. False
Most of the ribosomal RNA (rRNA) is synthesized in the nucleolus. Other RNA species, including 5S rRNA are synthesized outside the nucleolar region of the nucleus.

5. False
'Nucleosomes' are the repeating histone-DNA units that give eukaryotic chromatin its 'beads-on-a-string' appearance. Prokaryotic DNA is not associated with histones nor bounded by a nuclear envelope.

1.9 The eukaryotic (80S) ribosome:

1. dissociates into subunits with sedimentation coefficients of 50S and 30S.
2. is a multi-enzyme complex composed of RNA and protein.
3. contains two binding sites for transfer RNA (tRNA) molecules.
4. is primarily assembled in the nucleus.
5. dissociates into its constituent subunits on completion of protein synthesis.

1. False
It is convenient and now conventional to refer to the 'size' of macromolecules and organelles such as ribosomes in terms of their sedimentation coefficients. The intact eukaryotic ribosome (80S) is composed of a large subunit (60S) and a small subunit (40S). It is the smaller prokaryotic ribosomes (70S) that are composed of 50S and 30S subunits. Because the sedimentation of a macromolecular particle (and hence its S value) depends on both size and hydro-dynamic shape, S values are not necessarily additive.

2. True
The eukaryotic ribosome is constructed of several sizes of ribosomal RNA (5S, 5.8S, 18S and 28S rRNA) as well as about 70 proteins. The intact ribosome contains the enzyme machinery required for protein synthesis and can therefore be thought of as a multi-enzyme complex.

3. True
When both sites are occupied, the P-site holds the growing peptide chain attached to tRNA (peptidyl-tRNA) whereas the A-site holds the next aminoacyl-tRNA ready for peptide bond formation. (The terms P and A refer to peptidyl and aminoacyl respectively, although it is often convenient to think of the A-site as the 'acceptor' site.)

4. True
Most of the ribosomal proteins are imported into the nucleus from the cytoplasm. Within the nucleolus they form a large complex with the 45S precursor of the major rRNA components. This large ribonucleo-protein complex is then processed to form the precursors of the large and small ribosomal subunits. Final maturation of these subunits occurs during transport from nucleus to cytoplasm but the primary assembly occurs within the nucleus.

5. True

The large and small subunits of an individual ribosome are not permanently associated throughout their lifetime. The cytosol holds 'pools' of subunits which associate together to form ribosomes only on mRNA molecules. When the mRNA species have been translated, the ribosomes dissociate into separate subunits again.

Section 2

AMINO ACIDS AND PEPTIDES

2.1 The amino acid of structure

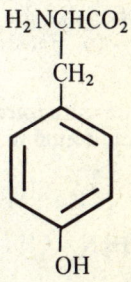

$$H_2NCHCO_2H$$
$$|$$
$$CH_2$$

OH

1. has only two ionizable groups.
2. would absorb light in the UV region near 280nm.
3. has a hydrophilic side chain (R-) group.
4. can act as a precursor of adrenaline (epinephrine).
5. is essential in the diet of humans.

1. False

The amino acid (tyrosine) has three ionizable groups, $-NH_2$, $-CO_2H$ and the phenolic $-OH$. However, the last of these is ionized only at relatively high pH (pH > 10).

2. True

All the aromatic amino acids, especially tryptophan and tyrosine, absorb in this region of the UV.

3. False

Because the phenolic $-OH$ is not ionized at physiological pH values, the predominant character of the side chain is that of the benzene ring, i.e. hydrophobic.

4. True

The conversion of tyrosine to adrenaline occurs mainly in the adrenal medulla and involves several steps.

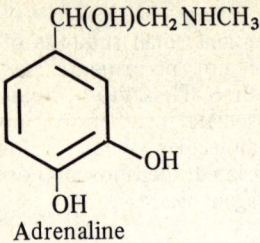

CH(OH)CH$_2$NHCH$_3$

OH

OH

Adrenaline

5. False

Tyrosine can be formed from phenylalanine and hence is not strictly an essential amino acid. However, it may become so if the supply of phenylalanine is limited or lacking. In the inborn error of metabolism, phenylketonuria, there is hereditary lack of the ability to transform phenylalanine into tyrosine.

2.2 The structures A–E are of five amino acids typically found in proteins.

(A)
$$CH_3$$
$$|$$
$$H_2N - CH - CO_2H$$

(B)
$$CH_3 \quad CH_3$$
$$\diagdown \diagup$$
$$CH$$
$$|$$
$$H_2N - CH - CO_2H$$

(C)
$$CH_2OH$$
$$|$$
$$H_2N - CH - CO_2H$$

(D)
$$CH_2S - CH_3$$
$$|$$
$$CH_2$$
$$|$$
$$H_2N - CH - CO_2H$$

(E)
$$CH_2 - CH_2$$
$$| \qquad |$$
$$CH_2 \quad CH - CO_2H$$
$$\diagdown \quad \diagup$$
$$N$$
$$H$$

These amino acids:

1. all have hydrophobic side chain (R-) groups.

2. all have non-ionizable side chain (R-) groups.
3. all can participate in α-helix formation in proteins.
4. are unable to form cross-links in proteins.
5. would be found in the L-form in proteins.

1. False
Amino acid C (serine) has a hydrophilic side chain: all the rest are hydrophobic.

2. True
The side chain of amino acid C (serine) is an aliphatic alcoholic group which does not ionize.

3. False
Amino acid E (proline) cannot form part of an α-helical region and is typically found at 'bends' in the polypeptide. All the others can fit into α-helical regions.

4. True
Although D (methionine) has a sulphur atom, this, unlike the -SH groups of cysteine, cannot form cross-links.

5. True
All of the amino acids in proteins have the L-configuration except glycine which is not optically active. However, D-amino acids do occur naturally in certain peptides, particularly in those found in bacterial cell walls.

2.3 The graph shows the titration curve for glycine hydrochloride.

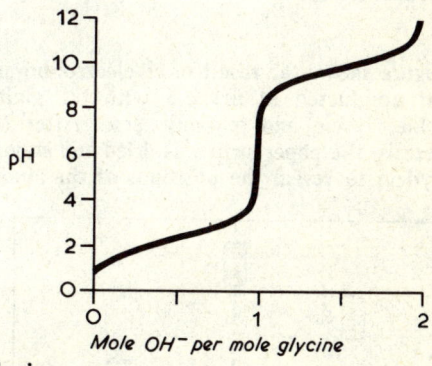

Glycine:

1. would act as a buffer over the pH range 1.3 to 3.3.

2. would act as a buffer over the pH range 4.3 to 6.3.
3. would act as a buffer over the pH range 8.6 to 10.6.
4. would carry zero net charge at pH 6.0.
5. would at pH 2.3, exist in equimolar amounts of $^+H_3NCH_2CO_2H$ and $^+H_3NCH_2CO_2^-$.

1. True
The 'buffering range' is identified as the pH range given by (pK ± 1). A pK can be identified from the curve at about pH 2.3 (the pH of half-neutralization). It is seen from the graph that adding (or subtracting) alkali near this point has relatively little effect on the pH.

2. False
In contrast to 1, addition of even a small quantity of alkali in this range will produce a dramatic change in pH.

3. True
See 1. There is another pK at pH 9.6, giving a second buffering range from 8.6 to 10.6.

4. True
This can be seen from the graph. Glycine-HCl has been treated with 1 equivalent of NaOH at this pH and will therefore be present in solution predominantly as $^+H_3NCH_2CO_2^-$.

5. True
At pH 2.3, the pK of the carboxyl group, glycine would exist predominantly in equimolar amounts of the two species shown.

2.4 The figure shows the result of an electrophoresis experiment conducted at pH 6.5 with the amino acids glycine, lysine and glutamic acid. After the electrophoresis, the paper strip was dried and stained with ninhydrin to reveal the positions of the amino acids.

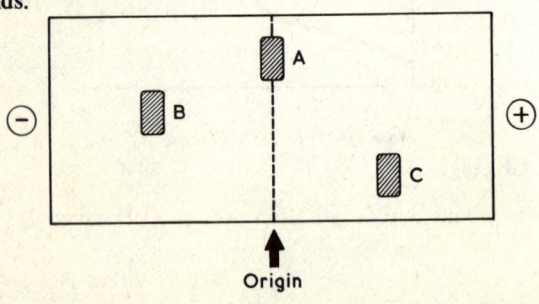

74

Which of the following deductions is or are true?

1. Amino acid C is glycine.
2. Amino acid B is lysine.
3. Amino acid A is glutamic acid.
4. Amino acid A has an isoelectric point near pH 6.5.
5. At pH 8, all the amino acids, A, B and C, would move towards the positive electrode.

1. False
Glycine will be virtually uncharged at pH 6.5 and would move little in an electric field.

2. True
At pH 6.5, lysine will have a net charge of +1 since the carboxyl group will have dissociated fully but the two amino groups will both carry positive charges. Lysine will therefore move to the cathode.

3. False
At pH 6.5, glutamic acid will have a net charge of −1 and will therefore move to the anode. Amino acid C is therefore glutamic acid.

4. True
At its isoelectric point, an amino acid has zero net charge and will not move in an electric field. At pH 6.5, amino acid A must therefore be quite close to its isoelectric point.

5. False
The addition of alkali effectively removes protons turning $-NH_3^+$ to $-NH_2$. This will increase the negative charge on A (glycine) and C (glutamic acid), but B (lysine) will still have a small positive charge and will still move to the cathode. Lysine is isoelectric at pH 9.7.

2.5 The following are sulphur-containing amino acids typically found in proteins:

$$
\begin{array}{cc}
\text{SH} & \text{CH}_2\text{S} - \text{CH}_3 \\
| & | \\
\text{CH}_2 & \text{CH}_2 \\
| & | \\
\text{H}_2\text{N} - \text{CH} - \text{CO}_2\text{H} & \text{H}_2\text{N} - \text{CH} - \text{CO}_2\text{H} \\
\text{(A)} & \text{(B)}
\end{array}
$$

Which of the following statements apply?

1. Amino acid A could form cross-links in proteins.
2. Amino acid B could form cross-links in proteins.
3. The keratin of feather, hair and nail contains relatively high amounts of amino acid A.
4. Amino acid B is an important donor of methyl groups in metabolism.
5. An enzyme that has amino acid A in its active site will be inhibited by treatment with sodium iodoacetate, ICH_2CO_2Na.

1. True
A is cysteine and could form the disulphide cystine $(R-S-S-R)$ with another cysteine residue in an adjacent stretch of polypeptide.

2. False
B is methionine and is a thioether, $R-S-CH_3$. This link is very stable (compare the methyl group in S-adenosylmethionine).

3. True
Keratins are rich in cysteine, which forms cross-links between chains. Permanent waving of hair involves breaking the links, curling the hair and then re-forming new links. The smell when feather and hair are burned also arises partly from the high sulphur content.

4. True
Methionine, by reacting with ATP, forms S-adenosyl-methionine which is then able to donate its methyl group to numerous compounds (e.g. to noradrenaline to give adrenaline).

5. True
Carboxymethylation takes place:

$$R-SH + ICH_2CO_2^- \rightarrow R-S-CH_2CO_2^- + HI.$$

Presuming the $-SH$ to be involved in enzyme activity, the activity will be lost when the covalent carboxy-methyl derivative is formed.

2.6 Consider the following compound:

$$CO_2H$$
$$|$$
$$S-S-CH_2CH-NH_2 \qquad CH_2CONH_2$$
$$| \qquad\qquad\qquad |$$
$$CH_2 \qquad\qquad CH_3 \qquad CH_2$$
$$| \qquad\qquad | \qquad\qquad |$$
$$H_2N-CH_2-CO-NH-CH-CO-NH-CH-CO-NH-CH-CO_2H$$

It contains:

1. four peptide bonds.
2. one amide bond in addition to the peptide bonds.
3. two disulphide bonds.
4. five amino acid residues.
5. glycine as its carboxy-terminal amino acid.

1. False
It contains three peptide bonds.

2. True
One amide ($-CONH_2$) is present.

3. False
It contains *one* disulphide bond.

4. True
Four amino acids are joined by peptide bonds. A fifth amino acid (cysteine) is linked by a disulphide bond. It could be stated that there are *four* amino acid residues present, one being cystine. However, this is not a useful concept. When proteins are being synthesized, cysteine units are incorporated, never cystine. Cystine forms later (post-translationally) by cross-linking of $-SH$ groups.

5. False
The carboxy-terminal amino acid is glutamine.

2.7 Consider the peptide:

Ala-Gly-Ser-Pro-Tyr-Lys-Met-Ala-Lys

This peptide was treated with dansyl chloride and then hydrolysed with 6M HCl at 110°C for four hours. Examination of the products of hydrolysis by thin-layer chromatography would be expected to reveal the presence of:

1. dansyl alanine.
2. monodansyl-lysine (N-ϵ-dansyl lysine).
3. bis-dansyl-lysine (N-α,ϵ-didansyl lysine).
4. O-dansyl serine.
5. O-dansyl tyrosine.

1. True
Dansyl chloride reacts with free amino groups: alanine is the *N*-terminal amino acid of this peptide.

2. True
Both the lysyl residues in the peptide will have free
ϵ-amino groups. Dansyl ϵ-N-lysine will therefore be
present in the acid hydrolysate.

3. False
The bis-dansyl derivative is only found where lysine
is the amino-terminal residue and thus has both its
amino groups available for dansylation.

4. False
The $-OH$ group of serine does not react with dansyl
chloride, and the dansyl derivative so formed is resist-
dansyl-serine) will only be obtained when serine is the
amino-terminal residue.

5. True
The phenolic $-OH$ of tyrosine reacts with dansyl
chloride and the dansyl derivative so formed is resistant to acid hydrolysis. Tyrosine, like lysine, will also
give a bis-derivative when it is the amino-terminal
residue of a peptide.

2.8 The peptide of structure:

$$\text{H}_2\text{N}-\text{CH}-\text{CONH}-\text{CH}-\text{CONH}-\text{CH}-\text{CO}_2\text{H}$$

1. contains valine as the N-terminal amino acid residue.
2. contains three peptide bonds.
3. would be expected to show hydrophobic properties.
4. would absorb light in the ultraviolet region at
   about 280nm.
5. would be cleaved by trypsin.

1. False
Valine is the C-terminal amino acid residue; the N-
terminal amino acid is alanine.

2. False
This is a *tri*-peptide and therefore has two peptide
bonds.

3. True
All the amino acid side-chains (R-groups) are either

aromatic or aliphatic with hydrophobic properties. The overall properties of the peptide would reflect this.

4. True
The presence of a phenylalanyl group would cause the peptide to absorb light in the UV region (280nm) but the absorption would be quite feeble. The other amino acids that absorb light in this region of the spectrum are tyrosine and tryptophan and the latter absorbs by far the most powerfully. Most proteins absorb light in this region of the spectrum because they contain Phe, Tyr and Trp residues.

5. False
Trypsin cleaves on the carboxy-side of arginyl and lysyl residues. The peptide *would* be cleaved by chymotrypsin which cleaves on the carboxy-side of hydrophobic residues.

Section 3

PROTEIN STRUCTURE AND FUNCTION

3.1 As applied to a protein, the term primary structure refers to:

1. the *number* of amino acid residues present in the protein.
2. the total amino acid *composition* of the protein.
3. the *sequence of amino acids* in the protein.
4. the α-*helical region* of the protein.
5. the *geometrical organization* of the polypeptide chain.

1. False
See 3.

2. False
See 3.

3. True
The amino acid sequence is referred to as the primary structure. Some authorities include the location of disulphide bridges as part of the primary structure.

4. False
α-Helical regions are 'secondary structure'.

5. False
See 3.

3.2 The following pattern was obtained following paper electrophoresis at pH 8.6 with a newly discovered haemoglobin variant, $Hb_{Toytown}$. Normal human haemoglobin, Hb_A, was electrophoresed simultaneously for comparison.

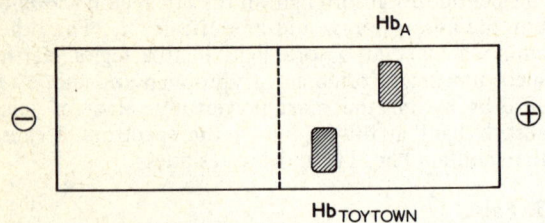

From the result of this electrophoresis experiment, it can be concluded that:

1. both haemoglobins have four subunits.
2. a mixture of the two haemoglobins could be separated from one another by ion-exchange chromatography on DEAE-cellulose.
3. the difference in electrophoretic mobility may result from a single amino acid change.
4. it is possible that a negatively charged amino acid residue in Hb_A has been replaced by a neutral one in $Hb_{Toytown}$.
5. it is possible that an uncharged amino acid residue in Hb_A has been replaced by a negatively charged one in $Hb_{Toytown}$.

1. False
They almost certainly have four subunits, but this cannot be concluded from this experiment.

2. True
Conditions could probably be found for this separation.

3. True
The difference in mobility may result from a single amino acid change but this experiment does not allow this to be concluded. Amino acid analysis and peptide mapping would be needed to establish this.

4. True
This is a possibility. Clearly, the overall negative charge on $Hb_{Toytown}$ is less than that on Hb_A because it has a lower mobility towards the anode. A definite answer would come from amino acid analysis and peptide mapping. A similar effect would be observed if a negatively charged amino acid were replaced by a positively charged one, or an uncharged amino acid by a positively charged one.

5. False
This is unlikely because it would give Hb$_{Toytown}$ a higher net negative charge. In this case, it would be expected to show a greater mobility towards the anode.

3.3 Two proteins, haemoglobin and lysozyme, have the following relative molecular masses (M_r values): **68 000** and **14 000**, and the following isoelectric points: **pH 6.8** and **pH 11.0**. A mixture of these two proteins could be separated by:

1. ion-exchange chromatography.
2. gel filtration.
3. dialysis.
4. electrophoresis at pH 7.0.
5. lyophilization.

1. True
The proteins have such widely different isoelectric points (pI values) that it would be easy to find an ion-exchange material and conditions such that the proteins could be separated.

2. True
The proteins differ sufficiently in molecular weight for this to be possible. Haemoglobin would be eluted first.

3. False
The pores in the normal dialysis membrane used in the laboratory hold back material of M_r greater than about 6000–8000. Both proteins would therefore stay in the dialysis bag.

4. True
Haemoglobin would hardly move at pH 7 but lysozyme would have a high mobility.

5. False
'Lyophilization' or 'freeze-drying' is a very gentle method of drying sensitive biological materials (e.g. proteins). The frozen solution is placed in a high vacuum. The water leaves by sublimation, leaving a dry, fluffy powder which stores well and rehydrates readily. Lyophilization is not a method for *separating* proteins.

3.4 The following result was obtained in an experiment in which immunoglobulin G (IgG) was subjected to electrophoresis in the presence of sodium dodecyl sulphate (SDS) and 2-mercaptoethanol. A set of standard reference proteins was electrophoresed simultaneously. The reference proteins were A, bovine serum albumin (M_r = 68 000); B, ovalbumin (M_r = 45 000); C, chymotrypsin (M_r = 25 000); D, cytochrome c (M_r = 12 000).

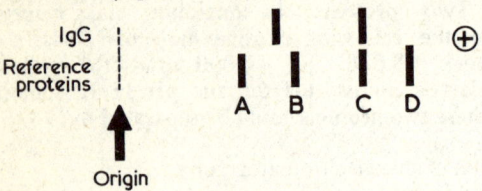

From these results it can be concluded that:

1. the M_r of IgG is 150 000.
2. IgG is made up of polypeptides of approximate M_r 25 000 and 50 000.
3. The M_r of IgG is likely to be 75 000 or some multiple of 75 000.
4. IgG is a glycoprotein.
5. the polypeptides of IgG are joined by disulphide bonds.

1. False
Although the M_r of IgG is indeed 150 000, this cannot be concluded from this experiment.

2. True
The position of the two bands of IgG, in comparison with the positions of the standards, allows it to be concluded that polypeptides of M_r 25 000 and 50 000 are present.

3. True
Because 2 is true, the M_r of IgG is highly likely to be $(25 000 + 50 000)_n$. (n in fact = 2 but it is not possible to conclude this from the experiment.)

4. False
Although IgG is indeed a glycoprotein, the result of this experiment gives no information about this.

5. False
Although the chains are joined by disulphide bonds, this experiment does not show that this is the case. Other experiments would be needed for this to be concluded, e.g. SDS gel electrophoresis in the *absence* of mercaptoethanol.

3.5 The structural protein collagen is found in tendon, bone and cartilage. In comparison with typical globular proteins, it has a number of 'unusual' features. These include:

1. a triple-helical structure.
2. the presence of the amino acid hydroxyproline.
3. the presence of a repeating unit $-(Gly-X-Y)-$ for the majority of its length.
4. an extremely low glycine content.
5. the presence of covalent cross-links between amino acid residues other than cysteine.

1. True
Collagen is composed of three polypeptide chains arranged as a triple helix. The α-helical structure found in many globular proteins is quite different.

2. True
This amino acid only occurs in collagen and a very few other proteins. It is formed from proline after incorporation into the polypeptide.

3. True
The unit is probably $-(Gly-X-Pro)-$ but the proline could also be hydroxyproline.

4. False
On the contrary, it has a relatively high glycine content (see 3).

5. True
Cross-links form between lysine residues in adjacent chains.

3.6 Different amino acid side chains (R-groups) give different regions of polypeptides their distinct properties.
With regard to protein structure:

1. typical globular proteins tend to have hydrophilic amino acid residues on the 'outside' exposed to the aqueous environment, and hydrophobic residues 'inside'.
2. proteins destined to be completely inserted into membranes would be expected to have hydrophobic amino acids on their surfaces.
3. some membrane proteins would be expected to have regions of hydrophobic and regions of hydrophilic amino acids on their surfaces.

4. the cleft into which the oxygen-binding haem group fits in the haemoglobin molecule provides a totally hydrophilic environment.
5. the cleft that forms the active site of the enzyme, lysozyme, and to which bacterial cell wall polysaccharide binds, provides a hydrophobic environment.

1. True
We know the complete structure of sufficient globular proteins to say that this is a general rule.

2. True
This is the opposite of 1: such proteins are 'designed' to exist in a hydrophobic environment.

3. True
Some membrane proteins have part of their bulk in the membrane phase (hydrophobic) and part exposed to the aqueous phase (hydrophilic).

4. False
The porphyrin ring (haem group) is almost insoluble in water and fits into a *hydrophobic* pocket in the globin. Haemoglobin variants in which just one of the amino acid residues lining the pocket is changed from hydrophobic to hydrophilic, tend not to bind haem. Such haemoglobins tend to bind O_2 non-cooperatively: usually only one pair of subunits is affected in this way.

5. False
The polysaccharide substrate is hydrophilic and will bind only to a hydrophilic cleft.

3.7 Which of the following structural proteins are extracellular?

1. Keratin.
2. Collagen.
3. Elastin.
4. Fibronectin.
5. Tubulin.

1. False
Keratin is an intracellular protein. For example, the dead cells of the *stratum corneum* of skin are

flattened and packed with keratin tonofilaments. Most epithelial cells have intracellular 'cytokeratins'. Even the hard keratins of hair are intracellular, each hair being made up of elongated or flattened, dead cells filled with keratin.

2. True
Collagen is extracellular and forms the cross-linked fibrous network of structural protein in, for example, connective tissue and cartilage. Glycosaminoglycan material forms the 'ground substance' in which the fibres are embedded to give a strong 'fibres-in-matrix' structure.

3. True
Elastin fibres exist as random coils and are extensively cross-linked. The resulting fibres or networks have great resilience. Connective tissue tends to contain mostly collagen with a small amount of elastin to provide flexibility and resilience.

4. True
Fibronectin is a fibre-forming protein which exists as large aggregates in the extracellular spaces. It promotes cell adhesion.

5. False
The structures known as microtubules (25nm diameter) are formed by the polymerization of the two types of subunit of the protein tubulin. Microtubules assemble only inside cells, where they form one component of the cytoskeleton. They also form the core of cilia and flagella in eukaryotic cells.

Section 4

ENZYMOLOGY

4.1 In an examination of the inhibition of an enzyme by a drug, I, the following Lineweaver-Burk plot was obtained when the reciprocal of initial velocity $(1/v)$ was plotted against the reciprocal of substrate concentration $(1/[S])$ for data obtained in the presence or absence of I.

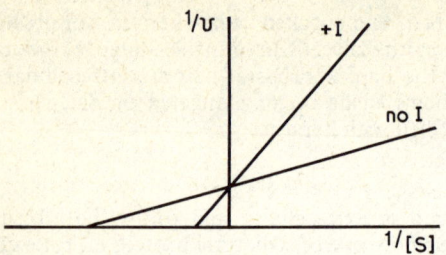

The experimental data obtained in this experiment would be consistent with:

1. irreversible inhibition of the enzyme by I.
2. competitive inhibition of the enzyme by I.
3. the type of inhibition observed when the reaction catalysed by succinate dehydrogenase is performed in the presence of malonate.
4. an increase in the apparent Michaelis constant (K_m) for the substrate.
5. negligible inhibition at very high substrate concentrations relative to the inhibitor.

1. False
Analysis of inhibition by the above method can only correctly be applied to *reversible* (non-covalent) interaction between enzyme and inhibitor. An example of an *irreversible* enzyme inhibitor is di-isopropylphosphofluoridate (Dip-F), which reacts covalently with the essential serine residue at the active site of acetylcholinesterase and the 'serine proteinases' (e.g. trypsin, chymotrypsin).

2. True
The common intersection point on the $1/v$ axis confirms competitive inhibition since it shows that a common maximum velocity (V_{max}) can be attained if the substrate concentration is high enough.

3. True
Malonate is sufficiently similar in structure to succinate for it to compete successfully for the substrate binding site.

4. True
Competitive inhibitors increase the apparent K_m since, in the presence of inhibitor, a larger concentration of substrate is required in order to reach V_{max}. This is confirmed by examining the intercepts on the $1/[S]$ axis).

5. True
At high concentrations of substrate (relative to inhibitor concentration), S tends to win the competition to gain access to the active site.

4.2 The reaction catalysed by chymotrypsin:

1. involves hydrolysis of a peptide bond on the C-terminal side of arginyl or lysyl residues.
2. takes place at a serine residue in the active site.
3. involves formation of a covalent acyl-enzyme intermediate.
4. has an optimal pH of 1.5–2.5.
5. is inhibited by di-isopropylphosphofluoridate (Dip-F or DFP).

1. False
This is the specificity of trypsin. Chymotrypsin is selective for peptide bonds on the C-terminal side of aromatic residues. Chymotrypsin can also hydrolyse esters such as p-nitrophenyl acetate (the acetyl ester of p-nitrophenol).

2. True
Chymotrypsin is one of the family of 'serine proteinases', which contain a serine residue in the active site.

3. True
Studies on the hydrolysis of p-nitrophenyl acetate by chymotrypsin provided the first evidence for the involvement of an acyl-enzyme intermediate in the catalytic mechanism.

4. False
Pepsin has an acidic pH optimum. Chymotrypsin has optimal activity near neutrality.

5. True
See 2. Serine proteinases are typically inhibited irreversibly by organophosphorus compounds such as Dip-F.

4.3 It is often convenient to assay an enzyme reaction by coupling it with another enzyme to form a product that is easily quantified. An example is the assay of pyruvate kinase by coupling it with lactate dehydrogenase (LDH):

$$\text{phosphoenolpyruvate} + \text{ADP} + \text{H}^+ \rightarrow \text{pyruvate} + \text{ATP}$$

$$\text{pyruvate} + \text{NADH} + \text{H}^+ \rightarrow \text{lactate} + \text{NAD}^+$$

In relation to this assay procedure:

1. 1mol NAD^+ is produced for each mol phospho-enolpyruvate metabolized.
2. of the two enzymes, pyruvate kinase should be in excess.
3. the reaction could be followed by measuring the decrease in absorbance at 340nm.
4. the reaction could be followed by measuring the decrease in absorbance at 260nm.
5. the total change in NADH concentration provides a measure of pyruvate kinase activity.

1. True
The stoichiometry should be obvious from the two reactions above.

2. False
For the assay to be valid, pyruvate kinase should be rate-limiting. Thus, *LDH* should be in considerable excess.

3. True
The oxidation of NADH may be followed at 340nm since NADH, but not NAD^+, absorbs at this wavelength.

4. False
Although adenine nucleotides absorb strongly at 260nm, there is no significant difference in the extinction coefficients for NADH and NAD^+ at this wavelength.

5. False
The total change in NADH concentration would provide a measure of the concentration of phospho-enolpyruvate or ADP, as long as NADH were in excess. It is the *initial rate* of NADH oxidation that provides a measure of pyruvate kinase activity.

4.4 Alcohol dehydrogenase catalyses the general reaction:

$$\text{RCH}_2\text{OH} + \text{NAD}^+ \rightleftharpoons \text{RCHO} + \text{NADH} + \text{H}^+$$

in which RCH_2OH can be almost any alcohol.

The Michaelis constant (K_m) of the enzyme for alcohol substrates:

1. is independent of the alcohol substrate oxidized.
2. is independent of enzyme concentration.
3. has the units of substrate concentration.
4. is equal to the affinity of the enzyme for its substrate.
5. would be unchanged in the presence of a non-competitive inhibitor.

1. False
The K_m can differ for different substrates used by the same enzyme. In the case of alcohol dehydrogenase a number of aromatic alcohols actually have much lower K_m values than ethanol itself.

2. True
Increasing the enzyme concentration will increase the initial rate of reaction but will not affect K_m.

3. True
The K_m corresponds to the substrate concentration that gives rise to half the maximum rate of reaction.

4. False
In many cases, K_m is approximately *inversely* proportional to the affinity of an enzyme for its substrate, but this is not a universal occurrence.

5. True
In contrast, a competitive inhibitor would cause an apparent increase in K_m. This is because a higher concentration of substrate is required to reach half-maximum velocity.

4.5 Lactate dehydrogenase from skeletal muscle is:

1. the rate-limiting enzyme in glycolysis.
2. located in the cytosol.
3. composed of four subunits.
4. an allosteric enzyme.
5. identical in subunit composition with lactate dehydrogenase from cardiac muscle.

1. False
Under most conditions, phosphofructokinase is the rate-limiting enzyme of glycolysis.

2. True
The glycolytic pathway is totally cytosolic. LDH is

frequently used as a 'marker enzyme' for the cytosol compartment in subcellular fractionation procedures.

3. True
Lactate dehydrogenase is a tetramer of subunits, each with M_r 35 000. There are two types of subunit (designated M and H), giving rise to five possible isoenzyme forms of LDH (M_4, M_3H, M_2H_2, MH_3, H_4). The subunits differ in charge and therefore the isoenzymes may be separated by electrophoresis. The M and H polypeptide chains are coded by separate genes which are expressed to different degrees in different tissues.

4. False
Not all oligomeric enzymes show allosteric properties.

5. False
The principal isoenzyme form in skeletal muscle and liver is M_4 whereas in heart it is H_4. The isoenzyme levels in serum are useful in diagnosis of heart and liver disease.

4.6 The Michaelis-Menten model of enzyme action:

1. assumes formation of an enzyme–substrate complex.
2. assumes formation of a covalent intermediate between enzyme and substrate.
3. explains the allosteric behaviour of certain regulatory enzymes.
4. explains the stereospecificity of enzyme reactions.
5. explains the maximum rate (V_{max}) attainable in enzyme-catalysed reactions.

1. True
The Michaelis-Menten model is the simplest that can account for the kinetic properties of many enzymes:

$$E + S \rightleftharpoons ES \rightarrow E + P$$

where E = enzyme; S = substrate; P = product.

The essential feature of the model is the formation of the enzyme–substrate complex, ES.

2. False
The enzyme–substrate complex may be covalent, but is of a transitory nature. The Michaelis-Menten model, though, assumes nothing about the nature of the binding of enzyme and substrate.

3. False
Allosteric enzymes do not generally obey Michaelis-Menten kinetics.

4. False
This is determined by the structure of the active site of an enzyme.

5. True
The maximum rate is achieved when all the enzyme active sites in solution are filled by substrate.

4.7 Allosteric enzymes, which typically control the rate of throughput in a metabolic pathway:

1. generally show a hyperbolic dependence of initial rate on substrate concentration.
2. are always oligomeric.
3. alter the equilibrium position of the reaction they catalyse.
4. catalyse the final reaction in a metabolic pathway.
5. are typified by aspartate transcarbamylase (ATCase) from *Escherichia coli*.

1. False
The most outstanding feature of allosteric enzymes is their non-hyperbolic (sigmoidal) dependence of initial rate on substrate concentration. Allosteric inhibitors increase the extent of the sigmoidal dependence whereas allosteric activators have the opposite effect.

2. True
There are some similarities with cooperative oxygen binding by haemoglobin. The binding of O_2 to haemoglobin (a tetramer) is sigmoidal whereas that to myoglobin (a monomer) is hyperbolic.

3. False
No enzyme or other catalyst can alter the equilibrium position of a reaction. Catalysts merely alter the rate at which equilibrium is attained.

4. False
Regulating the final reaction in a metabolic pathway is generally inefficient since it does not prevent the build-up of unwanted intermediates. Regulation generally occurs at an early step in a metabolic sequence (e.g. acetyl CoA carboxylase, aspartate transcarbamylase). This is referred to as 'feedback control'.

5. True
See 4. ATCase from *E. coli* regulates pyrimidine bio-synthesis and is inhibited in an allosteric manner by CTP. Note that, *in mammals*, the key regulatory enzyme in pyrimidine biosynthesis is carbamoyl phosphate synthetase which is inhibited by UTP and catalyses the step prior to ATCase.

4.8 Serum glutamate-oxaloacetate transaminase (GOT):

1. requires biotin as a cofactor.
2. activity is raised after myocardial infarction.
3. can be determined by adding suitable amounts of glutamate, oxaloacetate, lactate dehydrogenase (LDH) and NADH, and by measuring the decrease in absorbance at 340 nm.
4. activity would not be raised after liver damage.
5. is a source of ATP in erythrocytes.

1. False
Transaminases (aminotransferases) require pyridoxal phosphate, which forms a Schiff's base with the amino acid substrate.

2. True
Serum transaminase levels in normal subjects are low, but after tissue damage, these enzymes are released into serum. A specific example is heart muscle, and myocardial infarcts are accompanied by large rises in serum transaminase levels.

3. False
The reaction catalysed is:

$$\text{glutamate} + \text{oxaloacetate} \rightleftharpoons \alpha\text{-oxoglutarate} + \text{aspartate}$$

Neither of the products is a substrate for LDH and therefore no oxidation of NADH will occur. Trans-amination reactions involving alanine *can*, though, be measured by the inclusion of LDH and NADH in the coupled reaction system:

(i) alanine + α-ketoacid $\rightleftharpoons$ pyruvate + α-amino acid

(ii) pyruvate + NADH + H^+ $\rightleftharpoons$ lactate + NAD^+

4. False
Although glutamate-pyruvate transaminase (GPT) is a better indicator of liver damage than GOT, both transaminases are raised in acute hepatic disease.

5. False
No ATP is involved in the reaction.

4.9 The initial rate of an enzyme-catalysed reaction is:

1. normally directly proportional to substrate concentration.
2. normally directly proportional to enzyme concentration.
3. always maximal at neutral pH.
4. always maximal when the substrate concentration is equal to the K_m (Michaelis constant).
5. independent of temperature.

1. False
This is only true at very low substrate concentrations (relative to K_m). As the substrate concentration increases, the rate of reaction tends to a maximum value (V_{max}). The enzyme is then said to be 'saturated' with substrate.

2. True
Doubling the enzyme concentration will normally double the reaction rate, if all other conditions are held constant and components are in excess.

3. False
The pH value at which activity is at a maximum (the optimum pH) depends upon the particular enzyme. Enzymes normally have an optimum pH suited to their particular environment.

4. False
The K_m is the substrate concentration producing half the maximal velocity. Note that K_m has the units of concentration.

5. False
An increase of $10°C$ normally causes an approximate doubling of the rate of a chemical reaction. Enzymes, though, being proteins, will generally denature and begin to lose activity if the temperature is increased much above $35-40°C$. Thus, the effect of temperature on an enzyme-catalysed reaction is not simple to predict.

Section 5

CARBOHYDRATE STRUCTURE

5.1 Lactose:

1. is galactosyl-glucose.
2. is a non-reducing sugar.
3. is incapable of being digested by many human adults.
4. shows mutarotation.
5. cannot be digested by individuals with galactosaemia.

1. True
The galactose and glucose moieties are linked β-1,4.

2. False
Carbon-1 of the glucosyl residue is anomeric, i.e. it is an aldehyde carbon in the hemiacetal form. Therefore lactose is a reducing sugar.

3. True
Unless suffering from hereditary lactose intolerance, infants are capable of digesting lactose in milk. However, many adults in specific populations lose their ability to produce intestinal lactase. Thus, only 3% of Danes are deficient in lactase compared with 97% of Thais. Deficiency leads to an intolerance of milk, known as acquired lactose intolerance.

4. True
In solution, carbon-1 of the glucosyl moiety is in equilibrium with the open chain form. There are thus α and β forms of lactose.

5. False
Galactosaemia is an inborn error of metabolism in which there is a failure to convert dietary galactose to glucose. The accumulation of galactose in the blood has undesirable consequences including development of cataracts and mental disorders.

5.2 For aldohexoses (e.g. glucose), which have the formula $CHO(CHOH)_4 CH_2 OH$:

1. there are five asymmetric carbon atoms in the straight chain form.
2. there are 16 different molecular straight-chain

94

structures and therefore 16 chemically different sugars.
3. there are 16 different straight-chain structures and therefore 32 different ring structures.
4. D-glucose differs from L-glucose only in the configuration of groups on carbon-5.
5. all D-aldohexose sugars rotate plane-polarized light to the right.

1. False
An asymmetric carbon atom is one that has four different groups attached to it. There are only four asymmetric carbons in the straight chain form of aldohexoses.

2. False
There are 16 different molecular structures but eight of them are mirror images of the other eight. Therefore there are eight chemically different aldohexoses, each existing in the D and L form. Glucose, galactose and mannose are biologically by far the most important. Allose, altrose, gulose, idose and talose are only rarely encountered. The D-sugars are the naturally occurring forms.

3. True
When in the ring form, each of the eight D-sugars and eight L-sugars can exist as either the α or β structure. Effectively, there is an additional site of asymmetry at carbon-1. These α and β forms, which are said to be anomers of each other, are in equilibrium in aqueous solution. The spontaneous change of one to the other is known as mutarotation.

4. False
The D and L forms are non-superimposable mirror images of each other (i.e. they are stereoisomers, optical isomers or enantiomers). The D and L forms will therefore differ from each other by having the opposite configuration at each of the asymmetric centres.

5. False
A sugar is termed D or L by reference to the configuration at the asymmetric carbon furthest away from the carbonyl group and not by the direction in which it rotates plane-polarized light. The standard reference compound is glyceraldehyde, with D-glyceraldehyde being dextrorotatory. D-glucose (dextrose) is also dextrorotatory but the closely related D-fructose (laevulose) is laevorotatory.

5.3 'Reducing' sugars include:

1. glucose.
2. fructose.
3. sucrose.
4. galactose.
5. lactose.

1. True
A sugar is said to be 'reducing' if it is capable of reducing alkaline cupric sulphate to cuprous oxide (Fehling's test) or produces metallic silver from ammoniacal silver nitrate (silver mirror test). The reducing property is dependent upon the presence of an aldehyde group at carbon-1 when the sugar is in the open ring (straight chain) form. Glucose, an aldohexose, is therefore a reducing sugar.

2. True
Although fructose is a ketohexose it readily converts to glucose in weakly alkaline solutions and therefore reacts positively as a reducing sugar.

3. False
Sucrose is not a reducing sugar although it is a disaccharide consisting of glucose and fructose, both of which are themselves reducing sugars. The glycosidic bond is between carbon-1 of glucose and carbon-2 of fructose. These are the anomeric carbons of the two hexoses, respectively, so that no ring opening is possible for either of them and no aldehydic reducing group is formed.

4. True
Galactose differs from glucose only by the configuration of the hydroxyl group at carbon-4. It has an aldehyde group at carbon-1 when in the straight chain form and is therefore a reducing sugar.

5. True
Lactose is a disaccharide consisting of galactose and glucose. It is a β-galactoside, linked 1,4 to the glucose. The latter residue has potentially a free aldehyde group at carbon-1. It is therefore a reducing sugar.

5.4 Carbohydrate is stored as:

1. polyglucans in most living organisms.
2. cellulose in green plants.
3. lactose in the mammary gland.

4. glycogen in mammalian muscle as a source of blood glucose.
5. glycoproteins.

1. True
The commonest carbohydrate storage compounds are starch (amylose and amylopectin) in plants, and glycogen in animals. Amylopectin and glycogen consist exclusively of glucose residues, linked mainly α-1,4 but with some of them linked α-1,6 to produce branching. Amylose consists of α-1,4 linked glucose residues and is unbranched.

2. False
Cellulose is a structural rather than storage polysaccharide. As in many structural polysaccharides, the linkages between the sugar residues (in this case glucose) is β-1,4, rather than α-1,4.

3. False
Lactose is produced only when it is required for lactation. It is not a storage carbohydrate.

4. False
Glycogen is a storage polysaccharide in muscle but its function is to act as a source of energy for muscular contraction. It cannot give rise to blood glucose since muscle lacks the enzyme glucose 6-phosphatase.

5. False
Although glycoproteins contain anything from 1% carbohydrate to more than 30%, their function is not to act as carbohydrate storage compounds. Most proteins on the surface of animal cells are glycoproteins as are many secreted proteins. The blood-group antigens are also glycoproteins.

5.5 Glycogen:

1. is a macromolecule with a variable molecular weight.
2. has a highly branched structure.
3. is a reducing polysaccharide.
4. is the major source of stored energy in the body.
5. is deposited in excessively large amounts in the case of certain inherited metabolic diseases.

1. True
Although it is difficult to give an accurate figure for the molecular weight (particle weight) of fully

formed glycogen, it is probable that it normally ranges from 0.5 to 5 million, depending upon source. This corresponds to some 3000 to 30 000 glucosyl residues. As a macromolecule, glycogen makes no significant contribution to cellular osmotic pressure.

2. True
Glycogen consists of glucose units linked linearly α-1,4 with fewer residues linked α-1,6 to form branch points. Most glycogens branch on average about every 10–15 residues with the frequency depending to some extent on the source. Because the branching is repetitive half the molecule will be exterior to the final branch points, if the outer chains are fully formed. Being a highly branched polysaccharide, glycogen is soluble.

3. False
All of the outer chains of glycogen terminate in glucose residues attached to the rest of the molecule via the anomeric carbon-1; they are therefore non-reducing residues. There is only one glucose residue in a glycogen molecule that has a free hydroxyl group on carbon-1. Since this usually constitutes much less than 0.1% of the molecule, glycogen is non-reducing. The non-reducing ends of the outer chains (usually numbering many hundreds) are the metabolically active growth and deposition points of the glycogen molecule.

4. False
In an 'average person', carbohydrate (i.e. glycogen) constitutes less than 1% of the total fuel reserve. Triglycerides (neutral fat) of adipose tissue and protein (mainly in muscle) constitute about 80% and 20%, respectively, of the reserves that can be called upon during fasting and starvation.

5. True
A number of rare inborn errors of metabolism result in glycogen storage diseases (glycogenoses). Each results from a specific enzyme deficiency and causes glycogen, or glycogen with an abnormal structure, to accumulate.

5.6 Examples of important heteropolysaccharides are:

1. amylopectin.
2. heparin.

3. the polysaccharide component of murein (peptidoglycan).
4. keratin.
5. hyaluronic acid.

1. False
Amylopectin is a homopolysaccharide consisting only of glucose residues. It is similar to glycogen and forms the branched component of starch.

2. True
Heparin consists of alternating sulphated derivatives of N-acetyl glucosamine and iduronic acid (cf. glucuronic acid). It is a powerful naturally occurring inhibitor of blood-clotting. Heparin is used to prevent the clotting of blood samples taken by syringe and to inhibit clotting within blood vessels following heart attacks and other pathological conditions.

3. True
Murein (peptidoglycan) is the major structural component of most bacterial cell walls. It is a continuous lattice round the cell consisting of long polysaccharide chains of alternating β-1,4 linked N-acetyl glucosamine and N-acetyl muramic acid units. The polysaccharide chains are cross-linked by short polypeptide chains at each N-acetyl muramic acid residue.

4. False
Keratins are proteins found in hair, finger nails, wool, feathers and hooves. Keratan sulphate, which has no structural similarities with keratin, *is* a heteropolysaccharide found in cartilage, skin, bone and cornea.

5. True
Hyaluronic acid is a glycosaminoglycan (mucopolysaccharide) consisting of alternating units of glucuronic acid and N-acetylglucosamine. It is found in umbilical cord, synovial fluid, tendon and the vitreous humour of the eye. The glycosaminoglycans of connective tissue (chondroitin sulphate, keratan sulphate) are found in the form of proteoglycans which are polysaccharide complexes containing approximately 5% protein.

5.7 The common monosaccharides:

1. contain asymmetric centres.
2. are of two types, aldoses and ketoses.
3. tend to exist as ring structures in solution.

4. include glucose, galactose and raffinose.
5. are all readily formed from glucose by living cells.

1. True
A characteristic of sugars, with the exception of dihydroxyacetone, the simplest ketose, is that they contain one or more asymmetric carbon atoms. Each of these carbon atoms has four different groups attached to it, one of which is a hydroxyl in the unsubstituted monosaccharide.

2. True
Each monosaccharide in the straight chain configuration has a carbonyl group. This may be at carbon-1, where it is present as an aldehyde, or elsewhere in the chain, where it forms a ketone:

e.g.

$$
\begin{array}{ccc}
\text{CHO} & & \text{CH}_2\text{OH} \\
| & \text{or} & | \\
\text{CHOH} & & \text{C=O} \\
| & & | \\
| & & | \\
| & & |
\end{array}
$$

The two groups of monosaccharides are therefore respectively known as aldoses (or aldose sugars) and ketoses (or ketose sugars).

3. True
Monosaccharides with five or more carbons (pentoses upwards) normally exist as ring-closed structures in which the carbonyl group carbon has formed a linkage with a hydroxyl elsewhere in the chain. Rings with five and six atoms are preferred. These are known respectively as *furanose* and *pyranose* structures.

4. False
Raffinose is a plant trisaccharide consisting of galactose, glucose and fructose and is present in large amounts in sugar beet.

5. True
The common monosaccharides (trioses, erythrose, ribose, galactose, mannose, sedoheptulose, etc.) can all readily be formed from glucose by most organisms.

5.8 Polysaccharides:

1. contain many monosaccharide units which may or may not be of the same kind.

2. function mainly as storage or structural compounds.
3. may be easily separated from each other by ion-exchange chromatography or electrophoresis.
4. are present in large amounts in connective tissue.
5. may have amino acid residues attached after completion of the carbohydrate backbone, giving rise to glycoproteins and proteoglycans.

1. True
Polysaccharides are either homopolysaccharides or heteropolysaccharides depending upon whether the component monosaccharide units are of the same or different kinds. In heteropolysaccharides, the most common structure involves alternating units of two kinds of monosaccharide only.

2. True
Their main functions are to act either as carbohydrate storage compounds or as structural elements in cell walls or cell surfaces. Because they are so widespread in nature, they also serve an important nutritional role for many organisms.

3. False
Most polysaccharides are uncharged molecules at normal pH ranges and therefore do not lend themselves to separation by either ion-exchange chromatography or electrophoresis.

4. True
A number of polysaccharides such as hyaluronic acid, chondroitin sulphate and dermatan sulphate, known as glycosaminoglycans (or mucopolysaccharides), are present in connective tissue. Their role may be seen as structural in the sense that they form the matrix, or ground substance, in which fibres (e.g. collagen) are deposited, giving a strong, composite material.

5. False
Glycoproteins contain protein as the major component of the molecule whereas proteoglycans have carbohydrate as the principal component. In both cases, it is the carbohydrate residues, or relatively short chains of such residues, that are attached to the protein molecule, rather than the other way round.

5.9 Cellulose:

1. is the main structural component of cell walls of both green plants and bacteria.

2. is a linear polysaccharide consisting of 10 000 or more glucosyl units.
3. is in many respects similar to chitin, the structural material of the insect exoskeleton.
4. is a ready source of energy for all mammals.
5. is the most abundant naturally occurring macromolecule.

1. False
It is true that cellulose is the major structural component of cell walls of green plants. However, the material that confers the rigidity on bacterial cell walls is peptidoglycan, the polysaccharide component of which consists of N-acetylglucosamine and N-acetylmuramic acid.

2. True
Cellulose is a homopolymer of 10 000 or more glucose residues linked β-1,4. There is, therefore, no branching and the structure adopts an extended configuration. The individual molecules of cellulose are aligned side by side in insoluble bundles, in an almost crystalline organization, and are held together by intermolecular hydrogen bonding.

3. True
Chitin, like cellulose, is also a linear molecule with individual monosaccharide units linked by β-1,4 glycosidic bonds. The monosaccharide unit in this case is N-acetylglucosamine instead of glucose.

4. False
No enzymes capable of hydrolysing cellulose are secreted by mammals. Cellulose is therefore unavailable as an energy source for such animals unless, like ruminants and other herbivores, they carry in their digestive tracts large populations of microbial flora able to do this. In this case, the gut bacteria degrade or use the cellulose for their own purposes, eventually releasing compounds such as acetate and propionate which are absorbed by the mammal and form a major source of 'food'.

5. True
Cellulose is not only the most abundant macromolecule but, on a weight-for-weight basis, it is undoubtedly the most commonly occurring of all biological substances.

5.10 Consider the structure:

It is:

1. a sugar phosphate ester.
2. a phosphate anhydride.
3. a reducing sugar.
4. glucose 1-phosphate.
5. a ketohexose phosphate.

1. True
The phosphate (from phosphoric acid) is linked to a primary alcohol group ($-CH_2OH$) by an ester link.

2. False
Anhydrides are formed by the removal of a molecule of water between two acids. For example, pyrophosphate formed from two molecules of inorganic phosphate (phosphoric acid) is a phosphate anhydride. Similarly, the two terminal phosphate bonds in ATP are phosphate-anhydride bonds.

3. True
Because carbon-1 is unsubstituted, it is still in equilibrium (in the open chain form) with an aldehyde group. The compound is therefore a reducing sugar and can undergo mutarotation.

4. False
The ring numbering starts at the aldehydic carbon (see 3). This compound is therefore glucose 6-phosphate.

5. False
It is an aldohexose phosphate (see 4). Fructose 6-phosphate is an example of a ketohexose phosphate.

Section 6

CARBOHYDRATE METABOLISM

6.1 The pentose phosphate pathway leads to:

1. the complete oxidation of glucose to CO_2.

2. the generation of NADPH.
3. the production of glucose 1-phosphate for glycogen synthesis.
4. the production of free glycerol for triglyceride synthesis.
5. the production of 5-carbon sugars (e.g. ribose) for nucleic acid synthesis.

1. True
Operation of the pentose phosphate pathway will lead to complete oxidation of glucose without the involvement of the TCA cycle. Adipose tissue may oxidize much of its glucose (50% or more) by the pentose phosphate pathway.

2. True
The first two steps of the pathway are oxidations that generate NADPH, which is used for a number of important purposes especially fatty acid biosynthesis. The action of malic enzyme also provides NADPH for biosynthetic purposes:

$$malate + NADP^+ \rightarrow pyruvate + CO_2 + NADPH$$

3. False
Glucose 1-phosphate is not involved in the pentose phosphate pathway.

4. False
Free glycerol is not a product.

5. True
The pentose phosphate pathway is a major source of C_5 sugars.

6.2 Glycogen biosynthesis from glucose 6-phosphate in liver:

1. occurs by a reversal of the degradative (glycogen phosphorylase) pathway.
2. requires the participation of coenzyme A.
3. requires the participation of uridine diphosphate glucose (UDPG).
4. requires the participation of a branching enzyme.
5. is an energy-requiring reaction.

1. False
In vivo, glycogen biosynthesis occurs by a distinct pathway involving UDPG (see 3).

2. False
CoA is not involved in carbohydrate metabolism until acetyl CoA is formed from pyruvate.

3. True
The 'energy input' for glycogen synthesis, or the driving force, is the formation of UDPG from UTP and glucose 1-phosphate.

4. True
The addition of glucosyl units from UDPG, catalysed by the enzyme glycogen synthetase, produces only α-1,4 links. Another enzyme, branching enzyme, is required to form α-1,6 branch points.

5. True
The biosynthesis of glycogen does require energy, provided by UTP. However, the involvement of UTP and also the highly exergonic hydrolysis of pyrophosphate makes sure that the overall sequence is exergonic. In other words, it is 'driven' in the direction of biosynthesis.

6.3 The enzyme α-amylase:

1. hydrolyses glycogen to give glucose and maltose.
2. is present in saliva.
3. is activated by trypsin in the small intestine.
4. hydrolyses cellulose to glucose and maltose.
5. if absent from the pancreatic secretion, results in incomplete digestion of starch, a large bacterial population in the faeces and diarrhoea.

1. True
α-Amylase randomly hydrolyses the α-1,4 linkages of glycogen or starch to give a mixture of glucose and maltose; maltose is not hydrolysed. The α-1,6 linkages are not attacked and the residual core material is known as limit dextrin.

2. True
It is present in both saliva and pancreatic juice for the digestion of starch and glycogen.

3. False
It is not secreted as a zymogen and therefore does not require activation by proteolytic modification. α-Amylase requires calcium and chloride ions for full activity and has a pH optimum close to 7.

4. False

α-Amylase has no action on the β-1,4 linkages of cellulose. No mammalian enzymes exist for the hydrolysis of cellulose. Where it is a major component of the diet, it is hydrolysed by microbial action in the digestive tract, particularly in the rumen of ruminants or the caecum of other herbivores such as the rabbit.

5. True

Absence of pancreatic amylase is much more serious than absence of salivary amylase, indicating that the former plays the greater role in carbohydrate digestion. Absence of salivary amylase does not seem to be particularly disadvantageous.

6.4 The enzyme β-amylase:

1. hydrolyses the β-1,4 glycosidic linkages of cellulose.
2. hydrolyses starch to maltose.
3. is present in large amounts in certain germinating seedlings.
4. is present in saliva.
5. is present in mammalian lysosomes.

1. False

The enzyme is misleadingly named since it does not act on β-1,4 linkages, only on α-1,4. The terms α- and β-amylase refer to the two kinds of amylase rather than to the types of linkage hydrolysed.

2. True

β-Amylase attacks starch by removing successive maltose units from the non-reducing ends of the chains of α-1,4 linked glucose residues. In this respect, it differs from α-amylase which acts randomly to produce maltose and glucose.

3. True

It is responsible for the mobilization of stored starch in a number of germinating seedlings, particularly in barley and wheat. This is put to good use in the formation of malt in the brewing industry.

4. False

β-Amylase is not an enzyme of the mammalian digestive system.

5. False

The major carbohydrate-degrading enzyme of mammalian lysosomes is α-1,4 glucosidase (acid maltase).

6.5 The absorption of glucose in the digestive tract:

1. occurs in the small intestine.
2. is an energy-requiring process.
3. is stimulated by the hormone glucagon.
4. occurs more rapidly than the absorption of any other sugar.
5. is impaired in cases of diabetes mellitus.

1. True
Glucose and other sugars are absorbed by the epithelial cells of the brush border region of the small intestine.

2. True
Since glucose is usually transported against a concentration gradient by the epithelial cells of the small intestine, the process requires energy and is known as 'active transport'. Glucose transport is coupled to the inward transport of sodium ions down a concentration gradient. The energy is provided by the hydrolysis of ATP (by the Na^+/K^+ ATPase) which generates the Na^+ gradient by pumping Na^+ ions out of the cell.

3. False
Although glucagon is secreted by the α-cells of the pancreas in response to a low blood glucose concentration, it has no effect on glucose uptake by the small intestine. Its main function is to cause the activation of liver phosphorylase, thus leading to an increased rate of conversion of liver glycogen to blood glucose.

4. False
Galactose, which is also taken up by an active transport system, is absorbed faster than glucose.

5. False
There is no impairment of glucose uptake from the small intestine in cases of diabetes mellitus. Ingestion of glucose by the diabetic results in a rapid and large increase in blood glucose concentration (as is seen in the glucose tolerance test).

6.6 Disaccharides in the diet, or those formed from polysaccharide digestion:

1. are absorbed as such by the small intestine.

2. are hydrolysed to constituent monosaccharides by enzymes present in pancreatic juice.
3. are hydrolysed by specific enzymes of the small intestine.
4. are quantitatively unimportant since there are no enzyme deficiency diseases associated with their metabolism.
5. include lactose which, if not hydrolysed by β-galactosidase (lactase), causes galactosaemia.

1. False
They are first hydrolysed to their component mono-saccharides in the small intestine. However, since the hydrolytic enzymes are on the surface of the epithelial cells, absorption closely follows hydrolysis.

2. False
Although digestion takes place in the small intestine, the relevant enzymes are not present in pancreatic juice (see 1).

3. True
Specific enzymes, lactase (β-galactosidase), sucrase (invertase), maltase and isomaltase, which hydrolyse lactose, sucrose, maltose and isomaltose, respectively, are present on the microvilli of cells of the small intestine. These enzymes are not secreted into the lumen of the small intestine.

4. False
A number of specific disaccharide intolerances occur if disaccharide digestion is incomplete. The most important of these is hereditary lactose intolerance which results from a genetic deficiency in lactase. It has serious complications in babies where the symptoms are severe diarrhoea and malnutrition. On a lactose-free diet, the symptoms rapidly disappear.

5. False
Although lactose is an important dietary disaccharide, impaired hydrolysis will lead to a decreased rather than increased concentration of galactose in the blood.

6.7 Glucuronic acid:

1. is formed from glucose by oxidation at carbon-1.
2. is important in the formation of certain detoxification products.
3. is present in a number of polysaccharides.
4. is involved in the excretion of bilirubin.
5. is an intermediate in the biosynthesis of vitamin C.

1. False

Glucuronic acid, and other hexuronic acids, are formed from the appropriate hexose by oxidation of the $-CH_2OH$ group of carbon-6 to $-COOH$. Oxidation of glucose at carbon-1 (the aldehyde group or anomeric carbon) gives gluconic acid (as in the glucose oxidase reaction).

2. True

Glucuronic acid is important in the detoxification and excretion of certain drugs and other foreign compounds through the formation of soluble glucuronides.

3. True

Hyaluronic acid, a mucopolysaccharide of the intercellular ground substance of animal cells, and found in large amounts in umbilical cord and the vitreous humour of the eye, consists of repeating glucuronic acid-N-acetylglucosamine units. Chondroitin sulphate, a mucopolysaccharide of cartilage and bone consists of alternating glucuronic acid-N-acetylgalactosamine units.

4. True

Bilirubin, a highly insoluble breakdown product of haem, is transported to the liver where it is made more soluble by the attachment of two glucuronic acid residues. The conjugate, bilirubin diglucuronide, is secreted in bile.

5. True

Glucuronate is an intermediate in the synthesis of ascorbate (vitamin C) by plants and animals that can make this compound. Primates (including humans) and guinea pigs are among the few animals unable to carry out the last step in the pathway.

6.8 Glucose 6-phosphate:

1. is formed from glucose and ATP in the reaction catalysed by hexokinase.
2. is hydrolysed by glucose 6-phosphatase to glucose and inorganic phosphate.
3. is a 'high energy' phosphate ester.
4. is transported by the blood from liver to all tissues of the body.
5. is excreted in large amounts by infants suffering from galactosaemia.

1. True

Hexokinase catalyses the formation of glucose 6-phosphate from glucose and ATP in most tissues.

Another enzyme, glucokinase, also catalyses the same reaction in liver. Glucokinase has a much higher K_m for glucose than does hexokinase. Its rate is therefore much more dependent upon the blood glucose concentration and its role is to initiate the conversion of 'excess' blood glucose to liver glycogen.

2. True
Glucose 6-phosphatase is the enzyme responsible for the formation of glucose in the liver either from mobilized glycogen stores (glucogenesis) or during gluconeogenesis. Glucose 6-phosphatase is virtually absent from muscle, consistent with the fact that muscle glycogen does not give rise to blood glucose.

3. False
The hydrolysis of glucose 6-phosphate to glucose and inorganic phosphate is a 'low energy' hydrolysis and the reaction cannot be coupled to drive other phosphorylations, unlike the hydrolysis of ATP or phosphoenolpyruvate.

4. False
The mobile form of carbohydrate in the mammalian body is glucose.

5. False
During galactosaemia, galactose and galactose 1-phosphate accumulate in the blood and galactose is excreted in the urine.

6.9 Phosphofructokinase (PFK), catalysing the formation of fructose 1,6-bisphosphate from fructose 6-phosphate:

1. readily catalyses the reverse reaction under physiological conditions.
2. is allosterically activated by ATP.
3. is allosterically inhibited by citrate.
4. is allosterically activated by AMP.
5. is absent from skeletal muscle.

1. False
Phosphofructokinase catalyses the transfer of the terminal phosphate group from ATP to fructose 6-phosphate to give fructose 1,6-bisphosphate. The equilibrium position is very much in favour of the formation of the bisphosphate. Like many kinase reactions (using ATP in the forward direction) its action is reversed by a phosphatase, in this case

fructose 1,6-bisphosphatase. Because of the two unidirectional reactions, the interconversion of fructose 6-phosphate and fructose 1,6-bisphosphate is an important control point of glycolysis and gluconeogenesis.

2. False
ATP, although a reactant, is also an allosteric *inhibitor* of phosphofructokinase and thus of glycolysis. This is a form of end-product inhibition since the main function of glycolysis is to supply energy in the form of ATP.

3. True
Citrate is an inhibitory regulator of phosphofructokinase. When citrate levels in the cell are high, the energy requirements can be met other than by the continued breakdown of glucose or glycogen. Citrate may also be seen as a product of glycolysis (formed from pyruvate via acetyl-CoA and oxaloacetate) so that this is another example of end-product control of the pathway.

4. True
A high AMP level normally means that there is a low ATP level. The so-called 'energy charge' of the cell is low. The activation of phosphofructokinase by AMP increases the rate of glycolysis and thus tends to raise the ATP levels. Another physiologically important activator of PFK (and inhibitor of fructose 1,6-bisphosphatase) is fructose 2,6-bisphosphate which is formed from fructose 6-phosphate by a kinase distinct from PFK. Fructose 2,6-bisphosphate, at micromolar concentrations, relieves the inhibition of PFK by ATP.

5. False
Phosphofructokinase is a key enzyme of glycolysis and is therefore an important enzyme in the utilization of muscle glycogen.

6.10 Uridine diphosphate glucose (UDPG):

1. is a coenzyme derived from a B vitamin.
2. is a substrate for the enzyme glycogen synthetase.
3. is an intermediate in the synthesis of starch and cellulose in plants.
4. is required for the normal metabolism of galactose.
5. can be formed directly from UDP-galactose.

1. False
UDPG is a nucleoside diphosphate, consisting of uracil-ribose-phosphate-phosphate (UDP) with glucose attached to the terminal phosphate group via an ester linkage at carbon-1. There is no vitamin nor vitamin derivative in UDPG.

2. True
UDPG is an intermediate in the formation of glycogen from glucose 1-phosphate. The latter reacts with UTP to give UDPG and pyrophosphate. Glycogen synthetase then catalyses the transfer of the glucosyl residue from UDPG to a non-reducing end of a growing glycogen chain. In this way, an α-1,4 linkage is formed.

3. False
The mechanism of starch synthesis is similar to that of the synthesis of the α-1,4 glucosidic linkages of glycogen except that in most plants the intermediate is ADPG (adenosine diphosphate glucose) rather than UDPG. Cellulose, which has β-1,4 linkages between glucosyl residues is synthesized via ADPG, CDPG or GDPG depending upon the plant species.

4. True
UDPG reacts with galactose 1-phosphate in a reaction catalysed by uridyl transferase to give UDP-galactose and glucose 1-phosphate. Galactosaemia is usually caused by a deficiency of this enzyme.

5. True
The enzyme UDP-glucose 4-epimerase catalyses the interconversion of UDP-glucose and UDP-galactose. Glucose and galactose differ from each other only in the configuration of the hydroxyl group at carbon-4 of the ring; they are 4-epimers of each other.

6.11 Mammalian glucose 6-phosphate dehydrogenase:

1. catalyses the oxidative conversion of glucose 6-phosphate to ribulose 5-phosphate.
2. is sometimes congenitally absent in humans, in which case there is a deposition of abnormal amounts of glycogen in the liver.
3. is highly specific for NAD^+ as the hydrogen acceptor.
4. is important in erythrocytes in preventing methaemoglobinaemia.
5. if absent from erythrocytes, results in primaquine sensitivity.

1. False

It is the first enzyme of the pentose-phosphate pathway and, as such, catalyses the oxidation of glucose 6-phosphate to 6-phosphogluconolactone. The ring is then opened by a second enzyme (a lactonase) and a third enzyme, 6-phosphogluconate dehydrogenase, completes the oxidation to ribulose 5-phosphate.

2. False

Glucose 6-phosphate dehydrogenase is sometimes congenitally absent but this does not give rise to a glycogen storage disease. It is a deficiency in glucose 6-phosphatase which leads to the deposition of abnormal amounts of liver glycogen.

3. False

Mammalian glucose 6-phosphate dehydrogenase and 6-phosphogluconate dehydrogenase are highly specific for $NADP^+$ as the hydrogen acceptor. The enzymes from some bacteria, though, use NAD^+.

4. True

The NADPH produced by the pentose phosphate pathway is important in keeping glutathione in the reduced (−SH) form via the action of the enzyme glutathione reductase. Reduced glutathione, by preferentially reacting with hydrogen peroxide, prevents the oxidation of the iron of haemoglobin from the ferrous Fe(II) to the ferric Fe(III) state, a condition known as methaemoglobinaemia.

5. True

NADPH, again acting via reduced glutathione, is required to maintain the physical integrity of the erythrocyte membrane. The red blood cells of human individuals lacking glucose 6-phosphate dehydrogenase tend to undergo haemolysis, a condition which is made more acute by certain drugs such as the antimalarial drug primaquine. These individuals are said to exhibit primaquine sensitivity.

6.12 Gluconeogenesis:

1. is the formation of glucose from glycogen.
2. takes place primarily in the liver.
3. may use acetyl CoA as a precursor.
4. is stimulated during starvation.
5. allows the resynthesis of glucose and glycogen from lactate after vigorous exercise.

1. False
The formation of glucose from glycogen by the liver is known as glucogenesis. Gluconeogenesis and glyconeogenesis refer to the synthesis of glucose and glycogen, respectively, from non-carbohydrate precursors such as alanine or lactate. The overall process is essentially the reversal of glycolysis but with important differences at points where certain steps are effectively irreversible under physiological conditions.

2. True
The liver is the main site of gluconeogenesis. The process does not take place to any significant extent in muscle because of the low activities of glucose 6-phosphatase and fructose 1,6-bisphosphatase.

3. False
Pyruvate cannot be synthesized from acetyl CoA and therefore gluconeogenesis cannot use acetyl CoA as a precursor. Consequently there can be no net synthesis of glucose (or glycogen) from fat. *Intermediates* of the TCA cycle, however, can act as substrates for gluconeogenesis since they can be quantitatively converted to oxaloacetate.

4. True
During starvation, stored lipids are mobilized to provide energy. Tissue protein is broken down primarily to maintain an adequate concentration of blood glucose which is an essential energy source for the brain. A key regulatory step in gluconeogenesis is the reaction catalysed by pyruvate carboxylase. This enzyme is activated by acetyl CoA, a product of fatty acid oxidation.

5. True
Most of the lactate that is formed in the muscles during vigorous exercise passes into the blood and then to the liver. Here the lactate is converted to glucose which is returned to the bloodstream or stored as glycogen depending upon the further energy requirements of the body.

Section 7

TERMINAL PATHWAYS OF OXIDATION AND BIOENERGETICS

7.1 In the Krebs tricarboxylic acid cycle (or citric acid cycle):

1. 4-carbon acids such as fumarate have a catalytic effect on the rate of oxidation of pyruvate via the cycle.
2. the addition of malonate brings the cycle to a halt.
3. the addition of fluoroacetate brings the cycle to a halt.
4. 38 molecules of ATP are produced for each molecule of acetyl CoA fed into the cycle.
5. four molecules of NADH are produced for each molecule of acetyl CoA fed into the cycle.

1. True
The rate of operation of the cycle depends on there being sufficient oxaloacetate to combine with acetyl CoA (from pyruvate) to form citrate, which initiates the cycle of reactions. Because oxaloacetate is recovered after each turn of the cycle, the addition of even small amounts of fumarate (or indeed any of the acids of the cycle) gives rise to more oxaloacetate and therefore stimulates the rate of utilization of pyruvate.

2. True
Malonate acts as a competitive inhibitor of the enzyme succinate dehydrogenase, a key enzyme of the cycle. The use of malonate was crucial to the discovery of the cyclic nature of the pathway by Krebs. He found that succinate accumulated even from fumarate or malate which *follow* succinate in the cycle.

3. True
Fluoroacetate combines (as fluoroacetyl CoA) with oxaloacetate to form fluorocitrate which, however, fails to be dealt with by subsequent enzymes of the cycle. It is a powerful inhibitor of aconitase. Because of this blockage (inhibition), the cycle stops and fluorocitrate accumulates in large amounts. Fluoroacetate is therefore a dangerous poison.

4. False
About 12 molecules of ATP are potentially produced by one turn of the cycle, arising from the oxidation of 3NADH and 1FADH$_2$ via the electron transport chain (3 + 3 + 3 + 2 ATP). In addition, one molecule of GTP is also produced which is energetically equivalent to one ATP, hence a total of '12 molecules of ATP'. It is the complete oxidation of one molecule of glucose via glycolysis and the TCA cycle that yields about 38 molecules of ATP.

5. False
Only three molecules of NADH are produced: one at

the stage of oxidative decarboxylation of isocitrate, one at the stage of oxidative decarboxylation of α-oxoglutarate, and one at the stage of oxidation of malate. In addition one molecule of $FADH_2$ is produced by the succinate dehydrogenase step.

7.2 In mitochondria, the energy needed for synthesizing ATP from ADP and inorganic phosphate (P_i) is provided by:

1. passing electrons from molecular oxygen to the substrates of respiration.
2. passing electrons from compounds such as malate and succinate to molecular oxygen.
3. dissipating a concentration gradient of protons across the inner mitochondrial membrane.
4. dissipating a concentration gradient of pyruvate across the inner mitochondrial membrane.
5. converting glucose into lactate.

1. False
Electrons tend to pass from reduced compounds *to* oxygen which acts as an electron sink.

2. True
Reduced substrates can supply H^+ and electrons in reactions catalysed by dehydrogenases where NAD^+ or FAD may be the initial hydrogen acceptor. The passage of the electrons through a chain of electron carriers of decreasing oxidation-reduction potential releases free energy which can be used to drive the energy-requiring synthesis of ATP from ADP and P_i.

3. True
Passage of electrons through the electron transport chain generates a gradient of protons (and charge) across the inner mitochondrial membrane. The energy stored in this gradient is used in the synthesis of ATP from ADP and P_i when protons flow back into the mitochondrion via the F_0F_1-ATPase of the mitochondrial inner membrane.

4. False
Pyruvate generated from glycolysis passes into the mitochondria and is converted into acetyl CoA which can feed into the TCA cycle. The more pyruvate available, the faster the potential ATP synthesis by the TCA cycle and electron transport chain. However, the enzyme pyruvate dehydrogenase which

converts pyruvate to acetyl CoA is *regulated*, being
turned off by high levels of ATP. Oxidation of
pyruvate ultimately supplies the energy for ATP
synthesis but this has nothing to do with the gradient
of pyruvate concentration.

5. False
The production of lactate from glucose by glycolysis
takes place extra-mitochondrially. The oxidation
states of glucose ($C_6H_{12}O_6$) and lactate ($C_3H_6O_3$)
are in any case identical so no *oxidative* release of
energy can have taken place. Some ATP is produced
during glycolysis by substrate-level phosphorylation.
Energy is only obtained oxidatively from glycolysis
if the pyruvate which is formed is not reduced to
lactate.

7.3 The cytochrome P_{450} system:

1. is involved in the hydroxylation of steroids.
2. is involved in the hydroxylation of phenylalanine
   to tyrosine.
3. is involved in the hydroxylation of prolyl residues
   in collagen.
4. requires NADH for activity.
5. is located in mitochondria in the adrenal cortex.

1. True
Steroid hydroxylation is accomplished by an electron
transport chain involving cytochrome P_{450} as the final
component. The chain also involves NADPH, a
flavoprotein and a non-haem iron protein. The
oxygen of the incorporated hydroxyl group comes
from O_2, the other oxygen atom appearing in water.
This is a mono-oxygenase reaction.

2. False
Hydroxylation of phenylalanine to tyrosine is also a
mono-oxygenase reaction requiring O_2. In this case,
though, the reductant is tetrahydrobiopterin. Regeneration of the reduced biopterin cofactor is accomplished
by a NADH-dependent reductase. An inherited
deficiency of this reductase, or of the enzymes
involved in biopterin biosynthesis, can also produce
symptoms of phenylketonuria even if phenylalanine
hydroxylase levels are normal.

3. False
Prolyl hydroxylation involves a di-oxygenase reaction.
One atom of O_2 appears in hydroxyprolyl residues in

collagen. The other oxygen atom is incorporated into α-oxoglutarate which is converted to succinate and CO_2. Ascorbate is the reductant in this process.

4. False
NADPH is required; see 1.

5. True
The P_{450} system is also present in endoplasmic reticulum in liver where it plays a very important role in detoxification mechanisms.

7.4 Catabolic pathways may operate to:

1. use up excess ATP.
2. provide reducing power (as NADPH).
3. provide energy (for example as ATP).
4. keep the organism warm.
5. provide low molecular weight compounds which serve as precursors for biosynthetic pathways.

1. False
Catabolic pathways will produce ATP rather than use it. Anabolic pathways, which involve synthesis, require energy and this is often provided in the form of ATP.

2. True
The formation of reduced coenzymes may be considered the primary function of catabolic pathways. This 'reducing power' can be used in biosynthetic pathways as is the case, for example, with the synthesis of fatty acids.

3. True
The major function of catabolic pathways is clearly the provision of energy, in a readily usable form, normally as ATP. Most of this ATP arises from the reoxidation of NADH via the electron transport system (oxidative phosphorylation). A pathway such as glycolysis has the capacity to provide ATP under both aerobic and anaerobic conditions since substrate-level phosphorylation occurs.

4. True
The production of heat is usually coincidental to the other functions of catabolism. However, human infants and other mammals born hairless have a specialized fatty tissue called brown fat in the upper back and neck. The function of this tissue is to

generate heat by the oxidation of fat. The mitochondria of brown fat are essentially uncoupled since they do not normally produce ATP.

5. True
An important function of catabolism is to break down macromolecules to their constituent parts which are then used in biosynthetic pathways. For example, proteins are hydrolysed to the level of amino acids and these are used for the synthesis of new protein as well as being broken down further. Similarly, the TCA cycle operates to provide precursors for many biosynthetic pathways as well as playing a central role in the provision of energy.

7.5 Mitochondria:

1. take up NADH from the cytosol and reoxidize it to NAD^+.
2. transport one molecule of ATP to the cytosol for every molecule of ADP taken up.
3. contain all of the succinate dehydrogenase present in a cell.
4. are freely permeable to NH_3 but not to NH_4^+ ions.
5. contain high levels of glucose 6-phosphate dehydrogenase.

1. False
Intact mitochondria are impermeable to NADH. The hydrogens from NADH enter mitochondria either by being passed to dihydroxyacetone phosphate or to oxaloacetate. Both the α-glycerophosphate and the malate so formed can be taken up by mitochondria and then, on the inside, be re-oxidized back to dihydroxyacetone phosphate and oxaloacetate respectively. These are returned directly or indirectly to the cytosol but the hydrogens removed in forming them pass into the electron transport system.

2. True
ATP and ADP do not diffuse freely or independently across the inner mitochondrial membrane but are exchanged for each other by a specific translocase. Thus, as a molecule of ATP leaves the mitochondrion, one molecule of ADP is taken up and is then available for subsequent phosphorylation. The translocase is strongly inhibited by the plant glycoside, atractyloside.

3. True
Succinate dehydrogenase is one of the enzymes involved in the terminal oxidation of metabolites and is a component of the TCA cycle. None is present elsewhere in the cell. Succinate dehydrogenase is used as a 'marker enzyme' in cell fractionation studies, indicating the presence of mitochondria.

4. True
Free ammonia (NH_3), unlike the NH_4^+ ion, is a neutral molecule and readily crosses mitochondrial or plasma membranes. Because of the high pK of the ammonium ion, only about 1% of total ammonia is present in the blood as NH_3. Ammonia is very toxic particularly to the brain, and probably exerts its effect by reacting with oxoglutarate in the mitochondria to form glutamate. In this way, oxoglutarate is removed from the TCA cycle and the operation of the latter is consequently impaired.

5. False
Glucose 6-phosphate dehydrogenase and other enzymes of the pentose phosphate pathway are found in the cytosol, not in mitochondria.

7.6 The phosphorylation of ADP to ATP by mitochondria oxidizing succinate:

1. can only take place if the mitochondrial membranes are intact.
2. is driven by chemical coupling to the alternate oxidation and reduction of cytochromes via 'high-energy' intermediates.
3. obtains the necessary energy from the establishment of a proton gradient.
4. is inhibited by valinomycin.
5. is inhibited by rotenone.

1. True
The coupling of phosphorylation to oxidation (electron transport) requires mitochondria with an intact inner membrane. If the mitochondria are ruptured, oxidation of the substrate may still occur but the energy is not captured in the form of ATP but is dissipated as heat.

2. False
Early theories of oxidative phosphorylation postulated the existence of 'high-energy' phosphorylated intermediates which could transfer their phosphate

to ADP to form ATP. Despite intensive studies over many years, no such intermediates have been detected and the chemical coupling theory has been abandoned.

3. True
The chemiosmotic mechanism for oxidative phosphorylation depends on electron transport 'pumping' hydrogen ions from the matrix of the mitochondrion to the cytosol. The energy contained in this electrochemical gradient of protons is used to drive the synthesis of ATP as the protons 'flow' back into the mitochondrion.

4. True
The antibiotic, valinomycin, is an 'ionophore' which forms a lipid-soluble complex with K^+. It effectively makes the inner mitochondrial membrane permeable to potassium ions so that the energy from electron transport is used in the accumulation of these ions rather than being available to drive phosphorylation.

5. False
Rotenone, an alkaloid from *Derris elliptica*, specifically inhibits electron transfer within the NADH-ubiquinone reductase complex. It does not interfere with the oxidation of succinate because the electrons from this substrate enter the electron transport system beyond this point. Rotenone would, though, prevent the oxidation of an NAD^+-linked substrate such as malate.

7.7 The uptake of oxygen by isolated, coupled mitochondria:

1. is dependent upon the presence of an oxidizable substrate such as succinate or malate.
2. is greatly stimulated by the addition of ADP.
3. is inhibited by puromycin.
4. is inhibited by 2,4-dinitrophenol.
5. with most substrates, results in the formation of six molecules of ATP for every molecule of oxygen consumed.

1. True
No oxygen can be taken up unless there is a flow of electrons down the electron transport chain; this is what 'coupled' means. These electrons come mainly from the oxidation of TCA cycle intermediates or from fatty acids.

2. True
If the mitochondria are coupled, then oxidation cannot take place unless phosphorylation of ADP to ATP is also occurring. The availability of ADP is usually rate-limiting in this process and hence the addition of ADP stimulates oxygen uptake.

3. False
The antibiotic, puromycin, is a specific inhibitor of protein synthesis and has no effect on oxidative phosphorylation.

4. False
Dinitrophenol (DNP) is an uncoupling agent. By making the inner mitochondrial membrane permeable to H^+, it uncouples oxidation from phosphorylation. Oxidation of the substrates (and oxygen uptake) will still occur but no ATP will be formed since the proton gradient is abolished. Any energy released appears as heat.

5. True
The transfer of 2H to NAD^+ and the subsequent oxidation of the NADH via the electron transport chain results in the utilization of one atom of oxygen and the formation of three molecules of ATP. The P:O ratio is said to be 3. For most substrates, i.e. those passing hydrogen to NAD^+, six molecules of ATP will therefore be formed per *molecule* of O_2 consumed. Succinate, which passes hydrogen to FAD, has a P:O ratio of 2 when oxidized to fumarate.

7.8 Creatine kinase (creatine phosphokinase; CPK):

1. functions in liver to degrade phosphocreatine (creatine phosphate).
2. catalyses the hydrolysis of phosphocreatine to creatine and inorganic phosphate.
3. produces ATP from phosphocreatine in muscle.
4. is necessary for the transport of creatine into muscle cells.
5. is released into the blood in large amounts, following muscle damage.

1. False
Creatine kinase is not an enzyme of creatine degradation; it is also of relatively low activity in liver.

2. False
Kinases do not catalyse the hydrolytic removal of phosphate groups.

3. True

Creatine kinase plays an important role in the energetics of muscle and other excitable tissue. Its function is to catalyse the formation of ATP from phosphocreatine. This latter compound serves as a temporary storage form of 'high-energy' phosphate allowing the ATP concentration to be kept at a constant, high level. In humans, the amount of ATP in muscle is sufficient for less than a second of action. The creatine kinase reaction is readily reversible and phosphocreatine is re-synthesized at rest. Compounds such as phosphocreatine (and phosphoarginine in invertebrates) are known as phosphagens.

4. False

Creatine kinase does not have any known role in the transport of creatine or phosphocreatine.

5. True

Creatine kinase, referred to clinically as CPK, is of very high activity in striated muscle and consequently muscle damage will lead to an increased level of the enzyme in the serum. It is useful in aiding diagnosis of heart infarction where the serum level of CPK reaches a maximum some 20–30 hours after the heart attack.

Section 8

LIPID METABOLISM

8.1 The naturally occurring compound of structure:

$$CH_3(CH_2)_4 (CH=CH.CH_2)_2(CH_2)_6 CO_2 H$$

1. is linolenic acid.
2. is essential in the diet of humans.
3. would have its double bonds in the *trans* configuration.
4. would have a melting point higher than that of the corresponding fully saturated C_{18} acid.
5. is a precursor of prostaglandins.

1. False

It is linoleic acid. Linolenic acid is C_{18} but has three double bonds:

$$CH_3 CH_2 (CH=CHCH_2)_3(CH_2)_6 CO_2 H$$

2. True
Linoleic acid is one of the essential fatty acids.

3. False
The configuration of double bonds in almost all naturally occurring fatty acids is *cis*.

4. False
Just the opposite. High melting points are characteristic of (i) long chain, and (ii) highly saturated compounds, cf. stearic acid (C_{18} saturated), m.pt. 69.6°C; oleic acid (C_{18}, 1 double bond), m.pt. 13.4°C; linoleic (C_{18}, 2 double bonds), m.pt. −5°C.

5. True
All of the so-called essential fatty acids (e.g. arachidonic, linoleic, linolenic) are prostaglandin precursors.

8.2 Which of the following statements is/are true of triglycerides (triacyl glycerols)?

1. One gram of anhydrous fat stores one-sixth as much energy as a gram of hydrated glycogen.
2. Hydrolysis in the presence of NaOH will yield 3mol of the sodium salt of the fatty acids ('soaps') and 1mol glycerol.
3. Hydrolysis by pancreatic lipase will yield 3mol of fatty acid and 1mol glycerol.
4. Hydrolysis by the enzyme phospholipase C will yield 3mol of fatty acid and 1mol glycerol.
5. The triacyl glycerol occurring in the cells of adipose tissue represents a mechanical and thermal insulator as well as a store of energy.

1. False
On the contrary, fats represent highly concentrated stores of energy.

2. True
This is how soap is manufactured industrially from animal fat.

3. True
The reaction is an enzyme-catalysed hydrolysis: the products are essentially the same as in 2.

4. False
Phospholipases, like most enzymes, are highly specific. They act on phospholipids but not on triacyl glycerols.

5. True
This is the 'bonus' of storing energy as fat. The subcutaneous adipose tissue has thermal and mechanical insulating properties. These properties are especially important in animals such as seals and whales.

8.3 The breakdown of the fatty acid

$$CH_3 CH_2 CH_2 | CH_2 CH_2 | CH_2 CH_2 | CH_2 CO_2 H$$

via the β-oxidation pathway would:

1. not occur unless the coenzyme A derivative was formed first.
2. yield four mol acetyl units only.
3. yield four mol acetyl units and one mol propionyl units.
4. yield three mol propionyl units.
5. yield three mol acetyl units and one mol propionyl units.

1. True
Fatty acids need to be 'activated' before β-oxidation can commence. 'Activation' means forming the CoA derivative and some energy has to be expended to achieve this.

2. False
This is a C_9 compound. It cannot yield an even number of C_2 units.

3. False
There are not enough carbons for this.

4. False
Although this is theoretically possible (right number of carbon atoms), β-oxidation occurs by 2-carbon steps. 3-Carbon units (propionyl) only result if the final unit is C_3 rather than C_2.

5. True
Three acetyl and one propionyl unit is correct. The propionyl unit (C_3) can combine with CO_2 to form a succinyl unit (C_4) which can be fed into the TCA cycle.

8.4 The following sequence shows the key reaction in the β-oxidation pathway for catabolism of fatty acids:

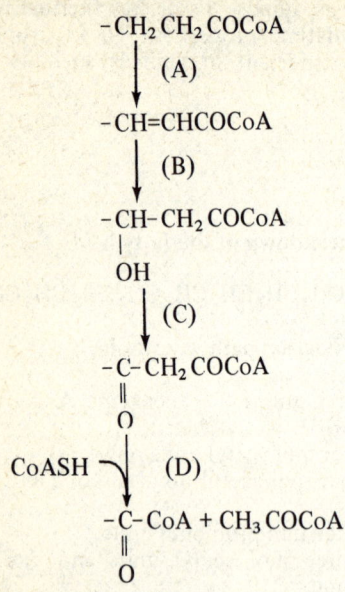

In this sequence of reactions:

1. steps A and C are dehydrogenations.
2. steps B and D are hydrolyses.
3. steps B and C are stereospecific.
4. step D requires energy in the form of ATP.
5. steps A–D take place in the cytosol of eukaryotic cells.

1. True
Step A involves FAD as a cofactor and Step B, NAD^+.

2. False
Step B is the addition of water catalysed by a hydratase: step D is a thiolysis, catalysed by a thiolase.

3. True
The third compound in the sequence, β-hydroxyacyl CoA, can exist in two forms because the β-carbon is optically active. The L-form is the one which occurs in this sequence. The enzymes producing it and using it are stereospecific.

4. False
Contrast this with the addition of CoA to the fatty acid $RCH_2 CO_2 H$, which does require ATP hydrolysis to drive it.

5. False
Fatty acid oxidation takes place in the mitochondria.

8.5 During β-oxidation of long chain fatty acyl CoA derivatives, energy is released. Which of the following best describe the way in which the energy is released and made available?

1. Both NADH and $FADH_2$ are produced which yield ATP via the electron transport chain.
2. NADPH is produced which yields ATP via the electron transport chain.
3. ATP is produced by substrate-level phosphorylation.
4. NADH and $FADH_2$ result from the entry of acetyl CoA into the TCA cycle and yield ATP via the electron transport chain.
5. Reversal of the reaction: fatty acid + CoASH + ATP → fatty acyl-CoA + AMP + PP_i, leads to the production of ATP.

1. True
Both NADH and $FADH_2$ are produced during β-oxidation. These can yield three and two ATP molecules respectively via the electron transport chain.

2. False
NADPH does not participate in β-oxidation. It is, however, involved in fatty acid *synthesis*.

3. False
No substrate-level phosphorylation occurs during β-oxidation.

4. True
Each C_2 unit of a long chain fatty acid becomes an acetyl CoA that can be oxidized via the TCA cycle.

5. False
This is the activation step, but the reverse reaction does not occur.

8.6 3-Hydroxy-3-methylglutaryl CoA (HMG-CoA):

1. is an example of one of the 'ketone bodies'.
2. is formed by the successive condensation of three molecules of acetyl CoA.
3. is an intermediate in the biosynthesis of cholesterol.

4. is an intermediate in the biosynthesis of palmitoyl CoA.
5. is formed in the degradation of leucine.

1. False
HMG-CoA is, though, an intermediate in the formation of the 'ketone bodies' (acetoacetate, 3-hydroxybutyrate and acetone) from acetyl CoA.

2. True
Two successive reactions are involved in the biosynthesis of HMG-CoA. First, 3-ketothiolase catalyses the condensation of two acetyl CoA to form acetoacetyl CoA. A further acetyl CoA then reacts with acetoacetyl CoA in the reaction catalysed by HMG-CoA synthetase.

3. True
The NADPH-dependent reduction of HMG-CoA to mevalonate catalysed by HMG-CoA reductase is the committed step in cholesterol biosynthesis. It is a key control point and dietary cholesterol leads to lowered levels of HMG-CoA reductase in the liver.

4. False
Fatty acid synthesis occurs through the addition of 2-carbon acetyl CoA units from *malonyl CoA*.

5. True
Leucine is a 'ketogenic' amino acid. It is degraded via HMG-CoA to acetyl CoA and acetoacetate.

8.7 The process of fatty acid biosynthesis shows several features which distinguish it clearly from fatty acid oxidation. These include:

1. its occurrence in the cytosol instead of in the mitochondria.
2. the requirement for ferredoxin.
3. the requirement for NADPH.
4. the participation of biotin.
5. the requirement for $FADH_2$.

1. True
Fatty acid biosynthesis occurs in the cytosol on a multienzyme complex, fatty acid synthetase.

2. False
Ferredoxin is a bacterial and plant, non-haem iron protein. It is not involved in fatty acid biosynthesis.

3. True
The reduction of $-CO-CH_2-$ to $-CH_2-CH_2-$ involves NADPH at two separate steps.

4. True
In the initial phase of fatty acid biosynthesis, CO_2 is added to acetyl CoA to produce malonyl CoA. The intermediate carrier of CO_2 is a protein-bound form of the B-group vitamin, biotin.

5. False
$FADH_2$ is formed during fatty acid *oxidation* but is not involved in fatty acid biosynthesis.

8.8 The fatty acid synthase complex of mammals contains:

1. covalently bound acyl carrier protein (ACP).
2. covalently bound pyridoxal phosphate.
3. seven pairs of identical subunits.
4. two identical subunits, each having several enzyme activities.
5. two different kinds of subunit, each having several enzyme activities.

1. True
It is a multienzyme complex with ACP covalently attached. ACP is a small polypeptide terminating in phosphopantetheine (cf. CoA).

2. False
Pyridoxal phosphate is not involved.

3. False
There are *two* subunits, see 4 and 5.

4. False
The subunits are different, see 5.

5. True
There are two subunits, each of which is coded by a single gene and each of which has a number of enzyme activities. Thus, subunit A contains acyl carrier protein, the condensing enzymes, and the β-ketoacyl reductase. Subunit B contains acetyl transacylase, malonyl transacylase, β-hydroxyacyl dehydratase and enoyl reductase.

8.9 The sequence of reactions below occurs, in the appropriate direction, in:

1. fatty acid biosynthesis.
2. fatty acid oxidation.
3. the TCA cycle.
4. glycolysis.
5. the pentose phosphate pathway.

$$-CH_2-CH_2- \xrightarrow{a} -CH=CH- \xrightarrow{b} -CHOH-CH_2- \xrightarrow{c} -CO-CH_2-$$

1. True
Once a new C_2 unit from malonyl CoA has been joined to the growing fatty acid chain, it is reduced, from $-COCH_2-$ to $-CH_2CH_2-$, by this sequence of reactions from right to left. NADPH is involved at steps c and a.

2. True
Left to right, this sequence represents the first three steps of the cycle (really a spiral) of reactions that form the basis of the β-oxidation pathway. $FADH_2$ is produced at step a and NADH at step c. Of course, the enzymes catalysing these reactions are different from those concerned with fatty acid synthesis.

3. True
Left to right, this sequence describes the steps between succinate and oxaloacetate in the TCA cycle. $FADH_2$ is formed at step a, and NADH is formed at step c as in β-oxidation. Of course, different enzymes catalyse the reaction compared with β-oxidation.

4 and 5. False
This sequence of reactions does not occur in either glycolysis or the pentose phosphate pathway.

8.10 Which of the following compounds is involved in the transport of long chain fatty acids into the mitochondrion across the inner mitochondrial membrane?

1. Oxaloacetate.
2. Citrate.
3. Carnitine.
4. Taurine.
5. Acyl carrier protein.

1. False
Oxaloacetate is involved in the transport of acetyl units out of the mitochondria (see 2).

2. False
Citrate is involved in the transport of acetyl units out of the mitochondria. Citrate is formed from acetyl CoA and oxaloacetate in the mitochondrial matrix. It then crosses the inner mitochondrial membrane and, once in the cytosol, reacts with CoA to reform oxaloacetate and acetyl CoA. This citrate cleavage reaction, but not the citrate synthase reaction, requires ATP.

3. True
Acyl carnitine, which is able to diffuse across the inner mitochondrial membrane, is formed from acyl CoA and carnitine. On the matrix side, the acyl unit is transferred back to CoA.

4. False
Taurine, an oxidation product of cysteine, does not take part in mitochondrial transport.

5. False
Acyl carrier protein is not involved in mitochondrial transport processes. It functions during fatty acid biosynthesis.

8.11 Fatty acid biosynthesis involves:

1. malonic acid.
2. malonyl CoA.
3. acyl carrier protein.
4. coenzyme A.
5. sulphydryl groups.

1. False
Free malonic acid is not involved in fatty acid biosynthesis (see 2).

2. True
Malonyl CoA is formed by combination of acetyl CoA and CO_2 carried by the biotin cofactor of fatty acid synthase.

3. True
Acyl carrier protein is a small protein (77 amino acids) which has a phosphopantetheine unit similar to that occurring in coenzyme A. The acyl carrier protein is attached to the fatty acid synthase complex.

4. True
In the initial stages of fatty acid biosynthesis, acetyl

CoA and malonyl CoA combine to form a C_4 compound, and CO_2 is released. Each time a C_2 unit is added, it comes from malonyl CoA.

5. True
The acyl carrier protein and coenzyme A bind the acyl unit via a sulphydryl group (see 3 and 4).

8.12 'Ketone bodies' (acetoacetate, β-hydroxybutyrate and acetone) are found in the blood of humans in large amounts:

1. immediately following a prolonged bout of vigorous exercise.
2. in untreated phenylketonuria.
3. in uncontrolled diabetes mellitus.
4. during starvation.
5. following a bout of deep breathing.

1. True
In prolonged exercise, when the muscle glycogen is used up, a switch to fat oxidation occurs. Partially oxidized fat is sent from the liver to other tissues, including muscle, as acetoacetate and β-hydroxybutyrate.

2. False
In this inborn error of metabolism, the 'ketones' that occur in blood and urine are phenylpyruvic acid and other products of the incomplete breakdown of phenylalanine. These are not related to 'ketone bodies'.

3. True
Diabetes mellitus has been referred to as 'starving in the midst of plenty' (see 4 below). Because of the lack of insulin, glucose, although plentiful, cannot enter the cells of the peripheral tissues which therefore experience glucose deficiency and switch to fat metabolism. Acetone, formed by non-enzymic decarboxylation of acetoacetate, can be detected on the breath of diabetics.

4. True
After the stores of glycogen have been used, the body is forced to switch to fat oxidation in order to obtain energy. One consequence of starvation (and diabetes) is the drawing of oxaloacetate from the TCA cycle to make glucose (by gluconeogenesis) to maintain the blood glucose level. Depletion of oxaloacetate means

that less acetyl CoA can be fed into the cycle just at a time when increased fatty acid breakdown is increasing the levels of acetyl CoA 'waiting' to enter the cycle. Condensation of acetyl CoA molecules to acetoacetate therefore occurs.

5. False
Hyperventilation does not cause ketosis.

Section 9

NITROGEN METABOLISM

9.1 Nitrogen fixation (conversion of N_2 to NH_4^+):

1. occurs in all plants.
2. requires the nitrogenase complex.
3. requires at least 12 ATP per N_2 fixed.
4. requires a powerful reductant.
5. requires the participation of CO_2.

1. False
Nitrogen fixation occurs only in prokaryotes (bacteria and blue-green algae). Some of the bacteria are symbiotic in the nodules on the roots of certain plants (e.g. legumes).

2. True
The nitrogenase complex contains the metals molybdenum and iron and is associated with a 'reductase' (Fe protein) which supplies electrons (see 4).

3. True
The $N{\equiv}N$ bond is very stable and is highly resistant to chemical attack. Quite a lot of energy has to be expended to fix nitrogen. [The chemical 'Haber process' uses an iron catalyst at $500°C$ and 300 atmospheres].

4. True
The reaction may be written:

$$N_2 + 6e^- + 12\,ATP + 12\,H_2O \rightarrow 2\,NH_4^+ + 12\,ADP + 12\,P_i + 4\,H^+$$

Electrons, usually from ferredoxin, pass to the reductase (see 2) which is an iron-sulphur protein. This then passes the electrons on to nitrogen on the nitrogenase complex.

5. False
Carbon dioxide is not involved in nitrogen fixation.

9.2 For mammals, some amino acids are essential in the diet, whereas others may be formed from dietary components. Humans are capable of converting:

1. arginine into lysine.
2. phenylalanine into tyrosine.
3. pyruvate into alanine.
4. aspartic acid into isoleucine.
5. oxaloacetate into aspartic acid.

1. False
This is not possible and lysine is an essential amino acid. Some plant proteins are poor in lysine and do not provide a balanced diet.

2. True
Phenylalanine can be converted to tyrosine in normal individuals by the enzyme phenylalanine hydroxylase. Tyrosine is not an essential amino acid provided there is sufficient phenylanine in the diet. Individuals with the inherited disease, phenylketonuria, are incapable of performing this conversion.

3. True
All amino acids can be replaced by the corresponding keto acid. Transamination, therefore, allows formation of alanine from pyruvic acid.

4. False
This is not possible. Isoleucine, like the other branched-chain amino acids leucine and valine, is an essential amino acid.

5. True
As in 3, a transamination is involved. Oxaloacetate can be 'drawn off' from the TCA cycle and converted to aspartate for the biosynthesis of substances such as pyrimidines.

9.3 Amino acids are metabolic precursors of a variety of important biomolecules. Mammals can convert:

1. tyrosine into thyroxine.
2. tryptophan into adrenaline.
3. histidine into histamine.
4. tryptophan into the nicotinamide ring of NAD^+.
5. serine into sphingosine.

1. True
This occurs in the thyroid gland. Iodine and two molecules of tyrosine are required. The process takes place not with the free amino acid but with tyrosyl residues in the protein thyroglobulin.

2. False
Tyrosine is the precursor of adrenaline.

3. True
Decarboxylation of histidine yields histamine.

4. True
The nicotinamide part of the NAD^+ molecule can be made from tryptophan. Thus, deficiencies in the vitamin, niacin, rarely result in the symptoms of pellagra unless the diet is also lacking in tryptophan.

5. True
Serine supplies the reactive terminal part of the sphingosine molecule.

9.4 The urea cycle:

1. supplies the bodily requirement for arginine in infants.
2. converts urea into uric acid.
3. converts ammonia into urea.
4. acts as an energy-supplying mechanism by oxidizing waste materials.
5. converts urea to ammonia and carbon dioxide.

1. False
In infants arginine is an essential amino acid and must be supplied in the diet. In adult humans, arginine is not essential since sufficient is synthesized by the urea cycle.

2. False
Uric acid is a product of purine degradation and has no connection with the urea cycle.

3. True
In mammals, the toxic ammonia, formed as a waste product from excess nitrogen in the diet, is converted to the much less toxic and highly soluble urea for excretion.

4. False
Quite the contrary; the body has to expend energy (ATP) to synthesize urea.

5. False
This is exactly the opposite of what the urea cycle achieves.

9.5 Which of the following compounds participate in or are closely associated with the urea cycle?

1. Ornithine.
2. Citrulline.
3. Isocitrate.
4. Arginino-succinate.
5. Phosphatidyl choline.

1. True
The urea cycle begins and ends with ornithine. (Arginase converts arginine to ornithine and urea in the last reaction of the cycle).

2. True
Citrulline is formed from ornithine and carbamoyl phosphate.

3. False
Isocitrate is an intermediate of the TCA cycle and although the TCA cycle has quite strong connections with the urea cycle these occur at the level of (a) oxaloacetate (as a precursor of aspartate) and (b) fumarate.

4. True
Arginino-succinate is formed when citrulline reacts with aspartate. The breakdown of arginino- succinate yields arginine and fumarate.

5. False
Phosphatidyl choline is a phospholipid and is not associated with the urea cycle.

9.6 Transamination (aminotransferase) reactions:

1. require the presence of coenzyme A.
2. convert one keto-acid to an amino acid whilst simultaneously converting another amino acid to its corresponding keto-acid.
3. involve the intermediate formation of a Schiff base.
4. will not operate properly in a vitamin B_6-deficient animal.
5. involve ATP hydrolysis to give AMP and pyro-phosphate (PP_i).

1. False
Coenzyme A is not involved in transamination reactions.

2. True
The transaminase reaction may be written:

keto-acid$_1$ + amino acid$_2$ ⇌ amino acid$_1$ + keto-acid$_2$

An example is:

pyruvate + aspartate ⇌ alanine + oxaloacetate.

3. True
A Schiff base intermediate is formed between the amino acid and the cofactor pyridoxal phosphate.

4. True
The cofactor required, pyridoxal phosphate, is synthesized from pyridoxine (vitamin B$_6$). The drug, isoniazid, which is used in the treatment of tuberculosis, produces vitamin B$_6$ deficiency by combining with pyridoxal phosphate to form a hydrazide.

5. False
ATP is not involved in transamination. It can be seen from the equations shown in 2, above, that there is little energy change in the course of such reactions.

9.7 The diagram shows a simplified form of the urea cycle.

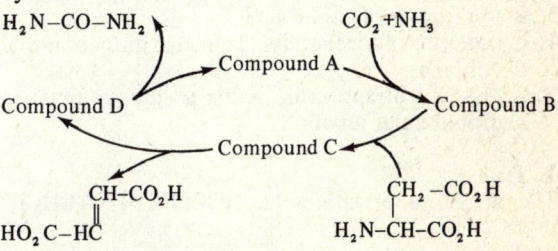

In this cycle:

1. compound A is arginine.
2. compound D is citrulline.
3. compound C is argininosuccinate.
4. the compound formed as a result of the conversion of compound C to D could enter the tricarboxylic acid cycle.
5. energy has to be supplied in the form of ATP.

1. False
Compound A is ornithine, formed when urea is produced from arginine.

2. False
Compound D is arginine, which loses its guanidino group in the formation of ornithine and urea (see 1 above).

3. True
Argininosuccinate is formed from aspartate and citrulline in a reaction that requires ATP.

4. True
The compound formed is fumarate, an intermediate of the tricarboxylic acid cycle.

5. True
The formation of 1 molecule of urea requires the hydrolysis of pyrophosphate bonds of ATP. Two molecules of ATP are required in the formation of carbamoyl phosphate from CO_2 and NH_3 which then reacts with ornithine (A) to form citrulline (B). One molecule of ATP is required in the formation of argininosuccinate and is hydrolysed to AMP and pyrophosphate.

9.8 Choline:

1. is a quaternary amine.
2. can function as a source of methyl groups.
3. is a precursor of cholic acid.
4. is present, as phosphatidyl choline, in mitochondrial membranes.
5. is the neurotransmitter acting at the mammalian neuromuscular junction.

1. True
The structure of choline is $HOCH_2CH_2N^+(CH_3)_3$.

2. True
Because the capacity of the body to produce methyl groups is limited, choline can be a dietary essential and has been classified as a vitamin. However, in the presence of adequate amounts of folic acid and vitamin B_{12}, it is not absolutely essential.

3. False
Cholic acid, one of the bile acids, is formed from cholesterol. It is unrelated to choline.

4. True
Phosphatidyl choline (lecithin) is one of several phospholipid components in eukaryotic membranes. In contrast, bacterial membranes are generally composed of one major type of phospholipid (phosphatidyl ethanolamine in *Escherichia coli*), and lack cholesterol.

5. False
Acetylcholine is the transmitter at the neuromuscular junction. It is formed by condensation of choline with acetyl CoA in the reaction catalysed by choline acetyltransferase.

Section 10

CONTROL OF METABOLISM

10.1 The activity of the enzyme adenylate cyclase can be potentiated by:

1. cholera toxin.
2. diphtheria toxin.
3. glucagon.
4. cortisol.
5. caffeine.

1. True
Cholera toxin activates adenylate cyclase irreversibly by catalysing the transfer of ADP-ribose from NAD^+ to the enzyme complex. This covalent modification (ADP-ribosylation) takes place on a GTP-binding protein associated with the complex.

2. False
Although there are considerable parallels in mechanism between cholera and diphtheria toxins, their sites of action are known to be quite different. Diphtheria toxin inhibits protein biosynthesis by inactivating the ribosomal translocase (elongation factor 2).

3. True
The principal hormonal actions of glucagon, stimulation of glycogenolysis or lipolysis, are mediated by cyclic AMP. Only those tissues with cell surface receptors for glucagon will respond to the hormone (e.g. liver but not muscle).

4. False
Cortisol, like all steroid hormones, mediates its effects by promoting the synthesis of specific mRNA species. The receptors for cortisol are intracellular and are not associated with adenylate cyclase.

5. False
Caffeine potentiates the action of cyclic AMP by inhibition of cyclic AMP phosphodiesterase. It does not affect adenylate cyclase itself.

10.2 Calcium ions, Ca^{2+}:

1. inhibit the action of the regulatory protein, calmodulin.
2. activate phosphorylase kinase.
3. activate the dependent- (D- or b-) form of glycogen synthetase.
4. are required for the blood clotting cascade.
5. are required for the release of noradrenaline from neurons.

1. False
Many of the biological actions of Ca^{2+} are mediated through calmodulin, an acidic protein of M_r 17 000. The Ca^{2+}-calmodulin complex activates cyclic AMP phosphodiesterase, phosphorylase kinase and many other enzyme systems.

2. True
Phosphorylase kinase can be activated by cyclic AMP-dependent phosphorylation or by Ca^{2+} (see 1). Activation by Ca^{2+} is particularly significant in muscle tissue where muscle contraction is initiated by the release of Ca^{2+} into the cytosol.

3. False
The D-form of the synthetase requires glucose 6-phosphate for activity.

4. True
Calcium ions are required at several stages in the conversion of fibrinogen to fibrin during clot formation. The chelation of Ca^{2+} by ethylenediamine tetra-acetate (EDTA), oxalate or citrate will prevent clot formation.

5. True
The uptake of Ca^{2+} into the presynaptic nerve terminal is needed for the release of neurotransmitter.

Calcium ions seem to be a general requirement for exocytosis to occur. Calmodulin is probably also required (see 1).

10.3 Lipolysis in adipose tissue:

1. is activated by adrenaline.
2. is activated by prostaglandin E1.
3. requires activation of lipoprotein lipase.
4. results in the release of glycerol into the blood-stream.
5. is inhibited by insulin.

1. True
Hormone-sensitive triacylglycerol lipase in adipose tissue is activated by cyclic AMP-dependent phosphorylation. The mechanism is analogous to that involved in regulating glycogenolysis in liver. A number of lipolytic hormones can activate lipolysis including adrenaline, glucagon and corticotropin (ACTH).

2. False
Prostaglandin E1 inhibits adenylate cyclase activity and therefore opposes the action of the lipolytic hormones.

3. False
Lipoprotein lipase (clearing factor lipase) is involved in the tissue uptake of triglycerides from circulating plasma lipoproteins.

4. True
Hydrolysis of triglycerides produces free fatty acids and glycerol. Adipose tissue lacks glycerol kinase and therefore the glycerol cannot be reutilized within the tissue.

5. True
Insulin has an antilipolytic effect. In diabetes mellitus, when insulin is lacking, lipolysis proceeds uncontrolled, ultimately producing large amounts of 'ketone bodies' which are excreted.

10.4 Citrate:

1. activates pyruvate carboxylase.
2. activates acetyl CoA carboxylase.
3. activates phosphofructokinase.

4. inhibits pyruvate dehydrogenase.
5. inhibits α-oxoglutarate dehydrogenase.

1. False
Pyruvate carboxylase is activated allosterically by acetyl CoA.

2. True
Acetyl CoA carboxylase is the key control site in fatty acid biosynthesis. When the levels of ATP and acetyl CoA are high, citrate levels increase which, in turn, activate the carboxylase.

3. False
The inhibitory effect of ATP on phosphofructokinase is enhanced by citrate. Thus, glycolysis is inhibited when biosynthetic precursors are plentiful. This is an example of 'feedback regulation'.

4. False
Pyruvate dehydrogenase, which regulates the production of acetyl CoA, is subject to multiple controls. Covalent modification by phosphorylation inactivates the enzyme. Dephosphorylation, which reactivates the enzyme is Ca^{2+}-dependent. Pyruvate dehydrogenase is also inhibited by the reaction products (NADH and acetyl CoA) as well as by ATP.

5. False
α-Oxoglutarate dehydrogenase is allosterically activated by Ca^{2+}, but is unaffected by citrate.

10.5 Fructose 1,6-bisphosphatase:

1. is a key regulatory enzyme in glycolysis.
2. is activated by ATP.
3. is activated by citrate.
4. is activated by cyclic AMP-dependent phosphorylation.
5. is inhibited by AMP.

1. False
Fructose bisphosphatase, catalysing the hydrolysis of fructose 1,6-bisphosphate to fructose 6-phosphate and inorganic phosphate, is a key enzyme in gluconeogenesis.

2. True
When ATP levels are high, fructose bisphosphatase is activated and gluconeogenic substrates are used to make glucose.

3. True
Contrast the effect of citrate on phosphofructo-kinase.

4. False
The enzyme is not regulated by covalent modification.

5. True
In fact the ATP/AMP ratio is the critical factor determining the activity of the enzyme.

10.6 The biosynthesis of many cell constituents such as amino acids, purines and pyrimidines:

1. normally begins with a pathway-specific enzyme that is essentially unidirectional.
2. is usually subject to control by negative-feedback inhibition.
3. in bacteria, is frequently regulated by induction of the necessary enzymes.
4. will decrease when ATP levels are high.
5. will invariably involve a reduction reaction at some stage in the pathway.

1. True
The first pathway-specific enzyme is normally uni-directional (with its equilibrium position well to the right) and rate-limiting.

2. True
The first enzyme, being rate-limiting and essentially irreversible, becomes an important regulatory point in the pathway. It is usual for such enzymes to be allo-steric and to be subject to negative-feedback inhibition by the end-product of the pathway. Regulation of the first enzyme of the pathway is very sensitive in that the activity will respond quickly to small changes in concentration of the end-product.

3. False
Biosynthetic pathways are normally regulated by repression of the synthesis of all the enzymes in the pathway. Such regulation is in addition to the allo-steric inhibition of the first enzyme of the pathway described in 2, above.

4. False
In general, high levels of ATP will favour biosynthesis whereas low levels of ATP (high ADP or AMP levels) will stimulate degradative pathways.

5. True

Just as degradation (catabolism) is associated at some stage with oxidation, so its reverse (biosynthesis) inevitably involves reduction. The reducing power may be provided as either NADH or NADPH but most frequently it is the latter that is used, as for example, in the synthesis of fatty acids.

10.7 The induction of β-galactosidase in *Escherichia coli*:

1. is under the control of the *gal* operon.
2. is brought about only by lactose.
3. is accompanied by the induction of a transport system for lactose.
4. is mediated through cyclic $3',5'$-AMP.
5. ceases if glucose is added to the growth medium.

1. False

The induction of β-galactosidase is under the control of the *lac* operon which is the genetic system coding for the proteins responsible for the uptake and degradation of lactose. The *gal* operon is the system concerned with uptake and degradation of galactose.

2. False

β-Galactosidase is induced when the cells utilize lactose. However, a number of analogues, particularly thiogalactosides such as methylthiogalactoside and isopropylthiogalactoside, act as gratuitous inducers in that they induce β-galactosidase but are not metabolized by it.

3. True

The structural genes of the *lac* operon code for β-galactosidase, a permease and a transacetylase; the function of the last of these is obscure. These genes are expressed co-ordinately in that, at constant temperature, their activities are in constant ratio to each other.

4. True

Before the promoter region of the operon can bind RNA polymerase, it must also bind a complex consisting of a protein (the catabolite activator protein) and cyclic AMP. The catabolite activator protein cannot bind to the promoter without first binding cyclic AMP. A number of other inducible enzyme systems require a similar indirect binding of cyclic AMP to the promoter before transcription can take place.

5. True
The addition of glucose to cells producing inducible, degradative enzymes, frequently 'switches off' the synthesis of such enzymes. This effect, originally known as the glucose effect, is part of the general phenomenon of catabolite repression. Glucose (and other catabolites) exert their effect by decreasing the level of cyclic AMP, thereby preventing the induction of unnecessary enzyme systems (see 4 above).

10.8 The change in the rate of enzyme synthesis occurring in many bacteria in response to changes in the nutritional environment:

1. is known as allosteric regulation.
2. is known as induction or repression.
3. is controlled at the level of translation.
4. frequently involves polycistronic messenger RNA.
5. is a mechanism which minimizes wasteful protein synthesis.

1. False
Allosteric regulation is the mechanism whereby the activity of a fixed amount of enzyme (in terms of number of enzyme molecules) is altered. It may consist of enzyme activation or inhibition.

2. True
The term enzyme induction is applied to an increase in the rate of specific enzyme synthesis. It is normally associated with degradative pathways where the increase in the amount of enzyme synthesized, when inducer (usually substrate) is added, may be many hundredfold compared with the activity when no inducer is present. Repression of enzyme synthesis refers to the 'switching off' of enzyme synthesis. It is normally associated with biosynthetic pathways where a build-up of end-product of the pathway usually exerts repression. Induction and repression of enzyme synthesis are most pronounced in bacteria, but they do occur to a lesser extent in mammalian systems.

3. False
The control of induction and repression takes place at the level of transcription. In each case, a repressor protein coded for by an operon-specific repressor gene is capable of binding to the operator region of the DNA and preventing the synthesis of mRNA (transcription). With inducible systems, the repressor

binds to the operator in the absence of inducer but not in its presence. With repressible systems, the repressor protein (or apo-repressor) can bind to the operator and prevent transcription only if it has first bound the end-product of the pathway (the co-repressor) or some substance related to it.

4. True
A polycistronic messenger RNA, i.e. one coding for more than one polypeptide chain and usually for more than one enzyme of the pathway, is normally associated with induction or repression.

5. True
Both enzyme induction and enzyme repression are mechanisms which prevent enzymes being synthesized unless they are required by the cell. The systems have added efficiency by being subject to operon control where related enzymes in a pathway are frequently induced or repressed together.

10.9 Phosphorylase kinase:

1. is converted to a catalytically active form on binding cyclic AMP.
2. is activated by phosphorylation of a specific tyrosine residue in the protein.
3. is activated by Ca^{2+} ions.
4. contains calmodulin as one of its constituent subunits.
5. activates glycogen phosphorylase by phosphorylation of a specific serine residue.

1. False
Cyclic AMP does not activate phosphorylase kinase directly. Cyclic AMP binds to the regulatory subunit of cyclic AMP-dependent protein kinase, releasing the catalytic subunit of the enzyme in an active form. Protein kinase then catalyses the phosphorylation of a range of proteins in the target cell including phosphorylase kinase (which is activated) and glycogen synthase (which is inactivated). These covalent modifications provide a co-ordinated control of glycogen synthesis and breakdown.

2. False
Cyclic AMP-dependent protein kinase catalyses the phosphorylation of specific serine residues in phosphorylase kinase. Protein kinases that catalyse the phosphorylation of tyrosine residues are also known but they are not involved in the regulation of glycogen metabolism.

3. True
Phosphorylase kinase can also be partially activated by Ca^{2+} ions which allows co-ordination of glycogen metabolism with muscle contraction.

4. True
Phosphorylase kinase is composed of four subunits, designated α, β, γ, and δ. The δ subunit is identical with calmodulin and is the site through which Ca^{2+} activates the enzyme. The γ polypeptide chain is the catalytic subunit and regulation of its activity is controlled by phosphorylation of the α and β subunits. Calmodulin also occurs free in the cell where, in conjunction with Ca^{2+}, it regulates the activity of other proteins.

5. True
The final step in the reaction 'cascade' that controls glycogen breakdown is the activation of glycogen phosphorylase by phosphorylase kinase. Again, phosphorylation occurs on a serine residue of the target protein.

Section 11

NUCLEOTIDE STRUCTURE AND FUNCTION

11.1 The methyl group of N^5-methyltetrahydrofolate (CH_3FH_4) can be directly donated in the:

1. conversion of homocysteine to methionine.
2. biosynthesis of phosphatidyl choline.
3. conversion of glycine to serine.
4. methylation of deoxyuridylate to form deoxythymidylate.
5. biosynthesis of the purine ring system.

1. True
CH_3FH_4 donates its methyl group to homocysteine in the reaction catalysed by methionine synthase. The reaction requires methylcobalamin (vitamin B_{12}) as a cofactor and takes advantage of the powerful nucleophilic properties of B_{12}. Homocysteine can also be methylated to methionine by other methyl donors such as betaine in the liver.

2. False
CH_3FH_4 is several orders of magnitude less reactive

as a methyl donor than a sulphonium compound such as S-adenosylmethionine. The latter is the principal methyl donor in the cell, and participates in the methylation of phosphatidyl ethanolamine to form phosphatidyl choline.

3. False
It is $N^{5,10}$-methylenetetrahydrofolate that participates in the conversion of glycine to serine in the reaction catalysed by serine hydroxymethyltransferase.

4. False
The 1-carbon donor in the synthesis of deoxythymidylate is $N^{5,10}$-methylenetetrahydrofolate. In the process, the folate derivative is converted to *dihydrofolate*. The active tetrahydrofolate cofactor is regenerated through the action of dihydrofolate reductase.

5. False
Carbon-2 of the purine ring is derived from N^{10}-formyltetrahydrofolate and carbon-8 is derived from $N^{5,10}$-methylenetetrahydrofolate.

methylidine

11.2 With respect to the urea cycle and pyrimidine biosynthesis:

1. carbamoyl phosphate is the source of one nitrogen atom and of the CO_2 utilized in urea synthesis.
2. carbamoyl phosphate and aspartate are the precursors required for the assembly of the pyrimidine ring.
3. the formation of carbamoyl phosphate used for pyrimidine biosynthesis occurs in the cytosol.
4. the nitrogen atom of carbamoyl phosphate used for pyrimidine biosynthesis is derived from glutamine.
5. the enzymes of the urea cycle are all located in mitochondria.

1. True
The other nitrogen atom of urea comes from aspartate.

2. True
The formation of carbamoyl aspartate from these precursors is the 'committed' step in pyrimidine biosynthesis.

3. True
Carbamoyl phosphate required for urea synthesis, however, is synthesized in mitochondria. *+ uses 2 ATP.*

4. True
However, when carbamoyl phosphate is synthesized for the urea cycle, the nitrogen atom is derived directly from NH_4^+.

5. False
Only the steps of the cycle up to citrulline synthesis occur in the mitochondrial matrix. The final reactions, including the synthesis of urea, occur in the cytosol.

11.3 Adenosine monophosphate (AMP):

1. is a nucleoside.
2. can be cyclized in a reaction catalysed by adenylate cyclase to give cyclic $3',5'$-AMP.
3. is a component of both RNA and DNA.
4. is an allosteric activator of phosphofructokinase.
5. is formed, with ATP, from two molecules of ADP by the action of myokinase (adenylate kinase).

1. False
AMP consisting of a base, a sugar and phosphate, is a nucleotide or a nucleoside monophosphate. A nucleoside (e.g. adenosine) consists of a base and a sugar only.

2. False
The substrate for adenylate cyclase is the 'high energy' triphosphate, ATP. This is cyclized with the elimination of pyrophosphate during the formation of cyclic AMP.

3. False
Adenosine monophosphate, since it contains ribose and not deoxyribose, is a component nucleotide of RNA but not DNA. However, it should be noted that synthesis of nucleic acids starts from the appropriate nucleoside triphosphates, not the monophosphates.

4. True
AMP is an allosteric activator of phosphofructokinase, a key control enzyme of glycolysis. AMP, therefore, indirectly stimulates the formation (or replenishment) of ATP by activating glycolysis.

5. True
Myokinase (adenylate kinase) catalyses the formation of one molecule each of ATP and AMP from two molecules of ADP. The enzyme is particularly important in muscle where it is closely associated with myosin ATPase in the myofibril and provides a mechanism for converting some of the ADP back

to ATP. The muscle cytoplasm (sarcoplasm) is rich in myokinase.

11.4 The hydrolysis of ATP to AMP occurs in the overall reaction catalysed by:

1. (Na^+/K^+)-ATPase.
2. acyl CoA synthetase.
3. adenylate cyclase.
4. argininosuccinate synthetase.
5. aminoacyl-tRNA synthetase.

1. False
In the ATPase reaction, ATP is hydrolysed to ADP and inorganic phosphate.

2. True
In the first stage of the acyl CoA synthetase reaction, a fatty acid reacts with ATP to form an acyl adenylate (acyl-AMP), releasing pyrophosphate. Then CoA displaces AMP from the acyl adenylate, forming acyl CoA.

3. False
Adenylate cyclase forms $3',5'$-cyclic AMP from ATP. Pyrophosphate is the other product.

4. True
The condensation of citrulline and aspartate to form argininosuccinate in the urea cycle is coupled to the hydrolysis of ATP to AMP.

5. True
The activation of an amino acid to aminoacyl-tRNA involves the formation of an aminoacyl-adenylate intermediate. The overall reaction is therefore analogous to that catalysed by acyl CoA synthetase. There is at least one aminoacyl-tRNA synthetase for each amino acid.

11.5 Guanosine triphosphate (GTP) is involved in the:

1. activation of adenylate cyclase by glucagon.
2. activation of Na^+ influx by acetylcholine at the neuromuscular junction.
3. elongation stage of protein biosynthesis.
4. reaction catalysed by phosphoenolpyruvate carboxykinase.

5. reaction catalysed by pyruvate carboxylase.

1. True
Activation of adenylate cyclase by hormones involves GTP. The hydrolysis of GTP to GDP deactivates the cyclase.

2. False
The binding of acetylcholine to its receptor at the neuromuscular junction leads to influx of sodium ions and depolarization of the cell membrane; there is no requirement for GTP at these *nicotinic* receptors for acetylcholine. Another class of acetylcholine receptors (muscarinic receptors) though, may involve GTP.

3. True
The elongation factor EF-Tu, which positions aminoacyl-tRNA species on the ribosome, contains bound GTP. During the elongation stage of protein biosynthesis, the GTP is hydrolysed to GDP.

4. True
The reaction catalysed is:

$$\text{oxaloacetate} + \text{GTP} \rightleftharpoons \text{phosphoenolpyruvate} + \text{GDP} + CO_2 .$$

5. False
The carboxylation of pyruvate requires ATP not GTP.

11.6 In humans, uric acid is:

1. the main end-product of purine metabolism.
2. present in excessive amounts in gout.
3. normally oxidized to form xanthine.
4. converted to urea in the liver.
5. excreted in larger amounts on a low-protein diet than during starvation.

1. True
The two major purines, adenine and guanine, are first converted to xanthine which is oxidized by the flavoprotein enzyme, xanthine oxidase, to hypoxanthine and then uric acid. Pyrimidines, e.g. thymine, are metabolized by a different pathway which involves their conversion to methylmalonyl CoA and then succinyl CoA.

2. True
Gout is initiated by an elevated level of uric acid

(principally in the form of sodium urate) in the serum. Urate is relatively insoluble and precipitates in the joints producing inflammation. A variety of biochemical lesions in purine metabolism can lead to excessive urate formation and hence produce gout. Inhibition of xanthine oxidase by allopurinol (an analogue of hypoxanthine) prevents the accumulation of urate and is used in the treatment of gout.

3. False
Just the opposite: xanthine is normally oxidized to form urate (see 1).

4. False
Urea is formed from urate in amphibians and most fish, but not in humans.

5. False
A low-protein diet does not lead to an increased production of uric acid. The majority of the purine content of the cell is normally reutilized via the 'salvage pathway' of purine metabolism rather than degraded to uric acid.

11.7 Mammalian ribonucleotide reductase:

1. catalyses the reduction of a ribonucleoside diphosphate (e.g. ADP) to the corresponding deoxyribonucleoside diphosphate (e.g. dADP).
2. uses NADPH as cofactor.
3. uses vitamin B_{12} as cofactor.
4. uses a protein cofactor such as thioredoxin.
5. is an allosteric enzyme.

1. True
Ribonucleotide reductase catalyses the formation of dCDP, dGDP and dADP. These nucleotides are then phosphorylated by a kinase to form their triphosphate derivatives, which are required for DNA synthesis. The reductase also catalyses the formation of dUDP from UDP. A series of reactions then converts dUDP to deoxythymidine triphosphate (dTTP) providing the fourth nucleotide needed for DNA synthesis. The key step in the formation of dTTP is the methylation of uracil (as dUMP) to form thymine (as dTMP) catalysed by thymidylate synthetase. Since ribonucleotide reductase is essential for the provision of the precursors for DNA synthesis, inhibitors of the enzyme (e.g. hydroxyurea) have been used in the treatment of cancer. Such inhibitors have limited

application, though, because of their toxicity to normal cells.

2. True
The overall reaction can be represented thus:

$$XDP + NADPH + H^+ \rightarrow dXDP + NADP^+ + H_2O$$

where $X = A, G, C$ or U.

3. False
Vitamin B_{12} is not a cofactor for *mammalian* ribonucleotide reductase although the enzyme from certain prokaryotic organisms does contain B_{12}.

4. True
Thioredoxin, a sulphydryl-containing protein of M_r 12 000 transfers electrons from NADPH to sulphydryl groups at the active site of ribonucleotide reductase. Thioredoxin may not be essential for ribonucleotide reductase activity since other sulphydryl reductants (e.g. glutaredoxin) can also function in this way.

5. True
There is a complex pattern of allosteric (feedback) regulation of ribonucleotide reductase which ensures that there is an adequate supply of all four deoxyribonucleotides for DNA synthesis.

11.8 Thymidylate synthetase:

1. catalyses the methylation of deoxyuridine triphosphate (dUTP) to form deoxythymidine triphosphate (dTTP).
2. uses methyltetrahydrofolate as methyl donor.
3. uses NADPH as cofactor.
4. uses a protein cofactor, thioredoxin.
5. is inhibited by fluorodeoxyuridylate (F-dUMP).

1. False
It is the nucleoside *monophosphate*, not the triphosphate, that is involved in the reaction catalysed by thymidylate synthetase.

2. False
The methyl donor in the reaction catalysed by thymidylate synthetase is 5,10-methylenetetrahydrofolate. In the process of the reaction dihydrofolate is formed, which is then reconverted to tetrahydrofolate

by dihydrofolate reductase. Methyltetrahydrofolate serves as methyl donor in the conversion of homocysteine to methionine. For most other methylation reactions S-adenosyl methionine is the methyl donor.

3. False
There is no involvement of NADPH in the reaction catalysed by thymidylate synthetase.

4. False
Thioredoxin regulates the activity of a number of proteins (e.g. ribonucleotide reductase) by serving as a sulphydryl reductant. However, it plays no role in the reaction catalysed by thymidylate synthetase.

5. True
F-dUMP, an analogue of the substrate dUMP, forms a covalent complex with thymidylate synthetase and the co-substrate 5,10-methylenetetrahydrofolate. However, the reaction can proceed no further and the enzyme is therefore 'locked' in an inactive state. F-dUMP is said to be a 'suicide inhibitor' of thymidylate synthetase. Inhibition of the enzyme ultimately prevents DNA synthesis and cell division and therefore F-dUMP has been used to a limited degree in cancer chemotherapy.

Section 12

NUCLEIC ACID STRUCTURE

12.1 In nucleic acids:

1. the normal nitrogenous bases in RNA are adenine, guanine, cytosine and uracil.
2. the bases in DNA are adenine, guanine, cytosine and thiamine.
3. the total purine content of native DNA is always equal to the total pyrimidine content.
4. the total purine content of RNA is always equal to the total pyrimidine content.
5. information for protein synthesis is stored as a triplet code in DNA.

1. True
The four common bases of RNA are A, G, C and U. However, modified bases, e.g. pseudouracil, are present in tRNA.

2. False
The nitrogenous bases of DNA are A, G, C and thymine, not thiamine. Thymine (T) is a pyrimidine base: thiamine is vitamin B_1 which is a completely different substance.

3. True
Native DNA is double-stranded and therefore base-paired. Since a purine always base-pairs with a pyrimidine (either adenine with thymine or guanine with cytosine) it follows that the total purine content must equal the total pyrimidine content.

4. False
RNA is single-stranded, except in certain viruses. Even though there may be considerable internal base-pairing, as in transfer RNA, it is incomplete and does not signify equality of pyrimidine and purine content.

5. True
The genetic information is, in most organisms, stored in DNA (the exception is certain viruses which contain only RNA). However, the genetic code as such is written as a triplet code in messenger RNA. Thus the codon for phenylalanine is written UUU and not AAA which would be the sequence of the complementary transcribed region of the DNA.

12.2 The two complementary strands of double-stranded DNA:

1. are held together by hydrophobic interactions.
2. are held together by hydrogen bonding.
3. will have the same amount of any one of the four bases in each strand.
4. may be separated from each other by gently heating in solution.
5. are held together with different strengths depending upon the source of the DNA.

1. False
The two strands are held together by stronger and more specific forces than are possible by hydrophobic interactions.

2. True
The two strands are held together by specific base-pairing between adenine and thymine and between cytosine and guanine (i.e. A=T; G≡C).

3. False
The two strands are complementary to each other, not identical. Therefore the amount of adenine in one strand will be equal to the amount of thymine in the other, and the same will apply to guanine in one strand and cytosine in the other. There is no requirement for any of the four bases to be present in equal amounts in the two strands.

4. True
The two strands of DNA will separate from each other if a solution of double-stranded DNA is heated. The separation temperature or transition point (also known as the 'melting point') is quite sharp and is accompanied by about a 40% increase in absorbance of the solution at 260nm. The process is reversible and 'annealing' takes place if the solution is cooled slowly.

5. True
The force with which the two complementary strands are held together is dependent upon the extent of hydrogen bonding. Since two such bonds are formed between each A–T pair, and three between a G–C pair, it follows that the higher the GC content then the more firmly the strands are held together. For example, the transition point (T_m) of DNA from *Escherichia coli* (50% GC) is 69°C whereas that from *Pseudomonas aeruginosa* (68% GC) is 76°C.

12.3 Transfer RNA (tRNA):

1. is responsible for transferring mRNA from the nucleus to the cytoplasm.
2. is usually about 150 nucleotides long.
3. is single-stranded.
4. contains many uncommon or modified bases.
5. is of many different types but always has the sequence CCA at the 3′ terminus.

1. False
The function of the different tRNA molecules is to carry amino acids to the site of protein synthesis. Each tRNA has an amino acid binding site and a codon recognition site. The attachment of the amino acid is catalysed by a specific aminoacyl-tRNA synthetase (or activating enzyme) which ensures that the correct amino acid is bound to the appropriate tRNA. The interaction between codon of the mRNA and anticodon of the aminoacyl-tRNA in

turn ensures that the amino acid is incorporated into the correct position in the growing polypeptide chain.

2. False
The tRNA molecules consist of between 73 and 93 nucleotides and therefore have M_r approximately 2500.

3. True
Each tRNA molecule is single-stranded but all show a high degree of internal base-pairing which results in the formation of double-stranded regions and produces the characteristic 'clover-leaf' structure.

4. True
A characteristic feature of tRNA is the high content of bases other than A, G, U and C. This varies roughly from 7 to 15 per molecule. The uncommon (or atypical) bases include such substances as hypoxanthine, pseudouracil, dihydrouracil and methylated derivatives of guanine and hypoxanthine. These modifications of the normal bases to form the atypical bases are carried out after assembly of the tRNA.

5. True
The sequences of over seventy different tRNA molecules are now known. The base sequence at the 3' end is always CCA and the amino acid is attached to the 3' hydroxyl of the terminal adenosine. The 5' end is phosphorylated and the terminal base is usually guanine.

12.4 The extra-chromosomal DNA present in many bacteria:

1. exists in the form of plasmids.
2. is single-stranded.
3. frequently carries genes that specify antibiotic resistance.
4. can often be passed between cells of different species.
5. may confer upon the cell the ability to take part in conjugation.

1. True
Most bacteria contain genetic elements in the form of relatively small DNA molecules that are free in the cytoplasm and which are separate from the large circular DNA chromosome. These additional DNA

molecules are known as plasmids. They are characterized by being unnecessary under certain conditions
and therefore often carried by only a few individual
cells in a bacterial population unless their presence
(and expression) is specifically required.

2. False
The plasmids are circular duplexes consisting of
double-stranded DNA and ranging in size from two
thousand to several thousand bases.

3. True
The plasmids often carry genes known as resistance
factors that code for the inactivation of specific
antibiotics.

4. True
One of the major characteristics of plasmids is that
they are transferable genetic elements that are not
just passed between cells of the same species but
which can, in many instances, be transferred between
different species.

5. True
Some plasmids, the fertility plasmids, enable bacteria
to transfer chromosomal DNA to each other during
direct cell to cell contact. This process is known as
conjugation and is made possible by the formation
of a conjugation tube between a so-called male
(donor) cell and female (recipient) cell. It is the
presence of the fertility plasmid, the F factor, which
confers 'maleness' on the donor cell.

12.5 Messenger RNA (mRNA):

1. is about 100 nucleotides long.
2. in bacteria is often long enough to code for more
   than one polypeptide chain.
3. will always have an AUG sequence near the 5' end.
4. is metabolically stable.
5. when transcribed from viral genes may code for
   more than one protein from the same nucleotide
   sequence.

1. False
The size of the mRNA will depend upon the protein
for which it codes and since there is a large range of
protein molecular weights there is a correspondingly
large range of mRNA sizes. A 'message' coding for a
polypeptide chain of M_r 30 000 will need a coding
sequence of at least 900 nucleotides.

2. True

A particular messenger RNA may often code for more than one polypeptide chain in cases where the polypeptides coded for have a functional relationship. The mRNA of the *lac* operon of *Escherichia coli* codes for three such proteins. That for the histidine biosynthesis system in *Salmonella* codes for nine enzymes involved in histidine biosynthesis. Such mRNA is said to be polycistronic.

3. True

AUG is the initiating codon for protein synthesis in both prokaryotic and eukaryotic organisms and since the message is read in the direction $5' \rightarrow 3'$, this codon will always be present near the $5'$ end. However, mRNA molecules are always somewhat larger than the polypeptide-coding region and the AUG codon may be preceded by a $5'$ untranslated sequence of 25 to 150 nucleotides.

4. False

Messenger RNA is normally synthesized and degraded very quickly, particularly in bacteria where the control of expression of certain genes is very sensitive and occurs rapidly in response to changes in the environment.

5. True

It is now known that a number of viruses, particularly bacteriophages, have regions of DNA and therefore of mRNA that code for up to three proteins (or parts of proteins) in the same nucleotide sequence. These genes are said to be overlapping. Each protein is specified by a different 'reading frame' on the mRNA.

12.6 Native DNA:

1. consists of two helical polynucleotide chains coiled round the same axis.
2. consists of two complementary polynucleotide strands.
3. has the sugar-phosphate repeat sequences on the inside of the structure forming the core of the strand.
4. in eukaryotic organisms, is usually found associated with large amounts of basic protein.
5. in prokaryotes, typically consists of one double-stranded circular chromosome.

1. True

The two chains are coiled round the same axis in a

right-handed double helix. The spatial arrangement between these two strands is such that there is a major and minor groove to the structure when viewed from the outside.

2. True
The two strands are complementary in that the base sequence of one strand specifies the base sequence of the other. This occurs because only certain pairs of bases are able to fit as a planar structure between the two individual sugar-phosphate chains. There are ten such base pairs for each turn of the double helix.

3. False
The base pairs are on the inside of the molecule, the sugar-phosphate sequences are therefore on the outside. This means that the majority of the charge on the molecule (the negatively charged phosphate groups) resides on the outside. The two chains are antiparallel (they show opposite polarity) in that the phosphate diester linkages between the deoxyribose units run $3' \to 5'$ in one strand and $5' \to 3'$ in the other.

4. True
Basic proteins, rich in arginine or lysine, and known as histones, are found in the chromatin of all nucleated eukaryotic cells. The histones interact with the negatively charged phosphate groups of the DNA to form bead-like nucleosomes along the length of the DNA. Other proteins, in addition to histones, also interact with DNA.

5. True
It is characteristic of prokaryotic cells that the main chromosomal material is arranged in a single, large, double-stranded DNA molecule. In *Escherichia coli*, this consists of about 4 million base pairs and has a contour length of approximately $1400\mu m$ compared with the $2-3\mu m$ length of the cell in which it is packaged.

12.7 Foreign DNA may enter a bacterial cell by the processes termed:

1. transcription.
2. transformation.
3. semi-conservative replication.
4. conjugation.
5. transduction.

1. False
Transcription is simply the process whereby RNA is synthesized alongside a template of DNA or in some cases alongside RNA (as in RNA viruses). It has nothing to do with the transfer of DNA between bacterial cells.

2. True
Transformation is the process whereby bacterial cells may take up and express foreign, naked DNA. It is the underlying process behind the classical experiment whereby dead, infective *Pneumococcus* may transfer virulence to a live but non-infective strain of the same organism.

3. False
Semi-conservative replication is the process whereby new copies of existing DNA are synthesized. It is semi-conservative in that one strand of the parent molecule is present in each of the two new duplexes that are formed.

4. True
Conjugation is the process whereby a 'male' or donor strain of a particular bacterium may take part in cell to cell contact with a 'female' or recipient strain and transfer some or all of its DNA to the recipient cell.

5. True
Infection of a bacterial cell by a DNA phage requires entry of the DNA into the cell. However, DNA from certain phages (the transducing phages) may also have associated with it some DNA from the previous host cell. The transfer and expression of this DNA is known as transduction.

Section 13

INFORMATION TRANSFER: FROM DNA TO PROTEIN

13.1 Several types of ribonucleic acid (RNA) can be found in cells. Of these:

1. ribosomal RNA (rRNA) is synthesized in the nucleolus.
2. messenger RNA (mRNA) is synthesized in the nucleus and carries the genetic information to the cytoplasm.

3. messenger RNA (mRNA) is the major type of RNA found in the cell.
4. transfer RNA (tRNA) acts as an 'adaptor' between the amino acid and the triplet code.
5. transfer RNA (tRNA) exists as at least 20 different kinds.

1. True
All the different types of RNA in the cell are synthesized in the nucleus, coded by base sequences in the DNA. Most of the rRNA genes are located in specialized regions of chromosomes that are associated with nucleoli.

2. True
mRNA is synthesized on the DNA of the nucleus: it subsequently passes into the cytoplasm to the sites of protein synthesis (the ribosomes).

3. False
Very little mRNA is found in cells: probably only 1–2% of the total RNA in cells is mRNA. Most of the RNA is rRNA representing about 80% of the total.

4. True
The genetic message exists as a sequence of bases in the mRNA as triplets of nucleotide bases. One end of the tRNA molecule can react with a specific amino acid and another part of the molecule can 'base-pair' with a triplet codon of the mRNA. Hence the 'message' is translated from DNA code into an amino acid sequence.

5. True
It follows from the above that for each type of amino acid there must be at least one type of tRNA molecule, corresponding at its other end to a triplet code. In fact, since there are several codes for some amino acids, there are actually many more than 20 kinds of tRNA.

13.2 The 'signal sequence' of secretory proteins:

1. specifies the point of attachment of carbohydrate residues.
2. directs the nascent protein into the endoplasmic reticulum.
3. determines the final conformation of the active protein.
4. is located at the *C*-terminus of the protein.

5. has a hydrophobic character.

1. False
N-Glycosylation of glycoproteins (on asparagine residues) occurs on the endoplasmic reticulum while translation is still occurring. Possible glycosylation sites are specified by the sequence Ser-Thr-X-Asn but not by the signal peptide. *O*-Glycosylation occurs post-translationally in the Golgi apparatus.

2. True
The function of the signal sequence is to direct newly synthesized secretory or membrane proteins into the cisternae of the endoplasmic reticulum.

3. False
The signal peptide sequence (typically 15–30 amino acids long) is normally rapidly removed and therefore has no role in determining the final active conformation of the protein.

4. False
Protein biosynthesis commences at the *N*-terminus. Since the signal sequence is at the *N*-terminus of the protein, this region of the message is translated first.

5. True
In all cases so far examined, the signal sequence has a relatively high proportion of amino acids of hydrophobic character.

13.3 Reverse transcriptase:

1. catalyses the synthesis of nucleic acids in the 3′ to 5′ direction.
2. is an enzyme coded by DNA tumour viruses.
3. can produce DNA from an RNA template.
4. can be used in the laboratory to synthesize complementary DNA (cDNA).
5. can produce RNA from a DNA template.

1. False
All known nucleic acid polymerases work in the 5′ to 3′ direction. 'Reverse' refers to the fact that this viral enzyme can make DNA from an RNA template, 'reversing' the more common flow of information from DNA to RNA to protein.

2. False
On the contrary, reverse transcriptase is a product of RNA tumour viruses such as the avian sarcoma virus.

RNA viruses containing reverse transcriptases are known as retroviruses.

3. True
The RNA of the virus is used as a template to make DNA after viral infection. Reverse transcriptase carries out this function.

4. True
Using reverse transcriptase and mRNA preparations as a template, cDNA can be synthesized.

5. False
Reverse transcriptase has DNA polymerase activity only.

13.4 The genome of a virus:

1. is known as a plasmid.
2. is always DNA.
3. may be integrated into the 'host' chromosome.
4. may comprise many (>100) genes.
5. may have overlapping genes.

1. False
Plasmids, found in bacteria and yeasts, are small circular DNA molecules which replicate independently of the main chromosome. They often carry genes such as those conferring resistance to antibiotics.

2. False
Viral genomes can consist of DNA (e.g. herpesvirus or the bacteriophage ØX174), or of RNA (tobacco mosaic virus or poliovirus).

3. True
Many bacterial viruses (bacteriophages) can form a prophage by integrating into the host chromosome and may remain inactive for long periods of time. They are said to be 'lysogenic', i.e. they have the potential to bring about lysis. Retroviruses are also capable of integrating into the host chromosome and this may be relevant to certain forms of cancer.

4. True
Some viruses have quite large genomes, e.g. T_4 bacteriophage has about 150 genes, many of which code for proteins required for synthesis and assembly of the intact phage.

5. True
This was a surprising discovery that emerged from
the determination of the entire sequence of the
genome of ØX174 (a small bacteriophage). Part of
the nucleotide sequence can code for two different
proteins, depending upon the 'reading frame'. Overlapping genes are also known for eukaryotic viruses.

**13.5 Several amino acid residues found in proteins
are formed as a result of post-translational events.
Such amino acids include:**

1. phosphoserine.
2. hydroxylysine.
3. γ-aminobutyrate.
4. glutamine.
5. γ-carboxyglutamate.

1. True
Protein phosphorylation, catalysed by protein kinases,
generally occurs on serine residues (e.g. glycogen
phosphorylase). Phosphorylation can also occur on
threonine and tyrosine residues.

2. True
Hydroxylysine is present in collagen and is formed
after lysine has been incorporated into the protein.

3. False
γ-Aminobutyrate is the main inhibitory neurotransmitter in the brain. It is formed by decarboxylation
of glutamate. It is not an α-amino acid and is not
found in proteins.

4. False
Glutamine is one of the 20 amino acids incorporated
into proteins ribosomally. The triplet codes for
glutamine are CAA and CAG.

5. True
γ-Carboxyglutamate is found in prothrombin as well
as in osteocalcin, a Ca^{2+}-binding protein in mineral-
izing tissue. γ-Carboxyglutamate is formed as a result
of a post-translational modification, catalysed by a
vitamin-K-dependent carboxylation system.

**13.6 Protein biosynthesis in both eukaryotes and
prokaryotes:**

1. begins with the *N*-terminus of the polypeptide.

2. occurs on 80S ribosomes.
3. is inhibited by puromycin.
4. is inhibited by cycloheximide.
5. uses AUG as the initiation codon.

1. True
Protein biosynthesis proceeds from the *N*- to the *C*-terminus. Methionine (or *N*-formyl methionine in prokaryotes) is the first amino acid to be inserted.

2. False
Prokaryotic ribosomes have a sedimentation co-efficient of 70S. Eukaryotic ribosomes are somewhat larger (80S). The ribosomes of mitochondria and chloroplasts, in contrast, seem to be more like those of prokaryotes.

3. True
Puromycin resembles the 3′-aminoacyl-adenosine end of aminoacyl-tRNA and can form a peptide bond with the growing peptide chain. This prevents further peptide bond formation and causes the release of unfinished polypeptides from the ribosome.

4. False
Cycloheximide inhibits peptide bond formation on eukaryotic ribosomes. Chloramphenicol has a similar action in prokaryotes.

5. True
The normal initiation codon is AUG, which codes for methionine. There may be many AUG codons in a mRNA but only one of these is recognized as the starting point for the peptide chain. Other parts of the mRNA sequence nearby identify the correct AUG for initiation. Mammalian mitochondria are unusual in that they can use AUA and AUU as initiation codons instead of AUG.

13.7 Histones:

1. are the major proteins present in chromatin.
2. are acidic proteins.
3. are coded in the DNA in multiple copies.
4. have amino acid sequences that have been highly conserved during evolution.
5. constitute the nuclear receptors for steroid hormones.

1. True
The mass of the histones in eukaryotic chromatin is

about the same as the mass of the DNA. Other proteins are present in smaller quantities.

2. False
Histones are basic proteins, rich in lysine and arginine.

3. True
The genes coding for the five major histones are clustered together and are repeated many times. The frequency with which these genes occur was deduced from the kinetics of hybridization of histone mRNA to the DNA. In human DNA, there are 30–40 copies of the histone genes.

4. True
Histones are among the most highly conserved proteins known. For example, there are only two amino acid differences in the sequence of histone H4 from pea seedlings compared with that from calf thymus.

5. False
The selective activation of genes by steroid hormones does not involve histones, but probably involves interaction with *non-histone* proteins in the chromatin.

13.8 Mitochondrial DNA:

1. normally exists in a circular double-stranded form.
2. is associated with histones.
3. codes for the rRNA of mitochondrial ribosomes.
4. codes for the mRNA species that specify the enzymes of the citric acid (TCA) cycle.
5. uses exactly the same genetic code as nuclear DNA.

1. True
Mitochondrial DNA, (with the exception of that of a few protozoa), is like bacterial DNA in this respect.

2. False
Only nuclear DNA is associated with histones and other proteins to form chromatin.

3. True
Although mitochondria have their own genetic material, they are not genetically self-sufficient. As the sperm contributes almost no cytoplasm to the fertilized egg, mitochondrial DNA is maternally inherited. The mitochondrial DNA of mammals is relatively small, containing about 16 000 base pairs, and codes for rRNA and tRNA species used by

mitochondria, as well as for a few polypeptide chains (see 4). The mitochondrial DNA has none of the extensive non-coding sequences seen in nuclear DNA.

4. False
The TCA cycle enzymes are coded by nuclear DNA and synthesized by cytoplasmic ribosomes. Mitochondrial DNA codes for cytochrome b as well as for some of the subunits of cytochrome oxidase and F_1-ATPase.

5. False
The concept that the genetic code is universal has had to be revised recently. There are a number of variations to the genetic code, e.g. AGA and AGG are read as stop signals rather than as arginine in mammalian mitochondria. The precise differences seem to depend upon the particular organism.

13.9 Both prokaryotic and eukaryotic types of mRNA:

1. are extensively modified (edited) before translation can occur.
2. are translated in the $5' \rightarrow 3'$ direction.
3. contain a sequence of a hundred or so adenine (A) residues at the $3'$ end.
4. are associated with a single ribosome at a time during translation.
5. can be polycistronic, i.e. can act as a template for the synthesis of more than one polypeptide chain.

1. False
Prokaryotic mRNA species undergo little or no modification after transcription. In fact transcription and translation tend to occur simultaneously. In contrast, eukaryotic mRNA species are initially synthesized as large precursors (heterogeneous nuclear RNA) which are extensively edited in the nucleus to form the final mRNA product which then proceeds to the cytoplasm for translation. The sequences of RNA that are removed and consequently not translated are called *introns*; the remaining sequences, which are then joined together and form the mature cytoplasmic mRNA are called *exons*.

2. True
mRNA is translated in the $5' \rightarrow 3'$ direction with the

polypeptide being synthesized starting at the *N*-terminus.

3. False
Most, but not all, eukaryotic mRNA species contain the 'polyA tail'; prokaryotic mRNA species lack the polyA tail.

4. False
A number of ribosomes may be translating a single mRNA molecule at any one time. Such a complex is called a polyribosome (*polysome*).

5. False
There is evidence for the existence of polycistronic mRNA only in the case of prokaryotes.

13.10 The replication of DNA:

1. occurs by a 'semi-conservative' mechanism.
2. requires the presence of all four deoxyribonucleoside 5′-diphosphates.
3. occurs only in the 5′ → 3′ direction.
4. requires the action of DNA ligase.
5. requires a primer of RNA.

1. True
The classical experiments of Meselson and Stahl demonstrated the semi-conservative nature of DNA replication. Both strands of DNA are copies, thus producing two new duplexes, each of which contains one of the original strands.

2. False
The deoxyribonucleoside triphosphates (dGTP, dCTP, dATP, dTTP) are required for biosynthesis. Inorganic pyrophosphate is released as each nucleotide is added to the growing polynucleotide chain.

3. True
All known DNA polymerases add nucleotides to the 3′-OH end of the growing polynucleotide chain.

4. True
Since synthesis can occur only in the 5′ → 3′ direction (see 3) and the two strands of the DNA run antiparallel, one of the daughter strands must be synthesized in small sections (Okazaki fragments). These are then joined together by DNA ligase.

5. True
A short stretch of RNA complementary to a section of one of the parent DNA strands serves to prime the synthesis of new DNA. This primer RNA is eventually removed and replaced by DNA.

13.11 The biosynthesis of RNA in eukaryotes is inhibited by:

1. actinomycin D.
2. rifamycin.
3. penicillin.
4. α-amanitin.
5. streptomycin.

1. True
Actinomycin D binds tightly to double-helical DNA and blocks the movement of RNA polymerase both in prokaryotes and eukaryotes. It is one of a class of compounds (ethidium bromide is another) that can insert between adjacent base pairs in the DNA, an interaction known as intercalation. Compounds of this kind, as a result of this intercalation, are also mutagenic.

2. False
Rifamycin, an antibiotic derived from *Streptomyces*, blocks the initiation of RNA synthesis in prokaryotes only, by binding to and inhibiting RNA polymerase.

3. False
The antibiotic penicillin inhibits bacterial cell wall biosynthesis and has nothing to do with RNA biosynthesis.

4. True
α-Amanitin comes from the 'death cap mushroom' (*Amanita phalloides*) and is extremely toxic. It particularly inhibits the RNA polymerase that makes mRNA in eukaryotes. The fungus also produces another toxin, *phalloidin*, that blocks the movement of actin filaments (microfilaments) in cells.

5. False
Streptomycin prevents initiation of protein synthesis on prokaryotic and other (70S) ribosomes. It can also cause the mRNA to be misread. Streptomycin is strain-specific and binds to a protein in the 30S subunit of 70S ribosomes in susceptible strains.

13.12 Bacterial restriction endonucleases:

1. can hydrolyse single-stranded DNA molecules.
2. usually hydrolyse DNA adjacent to 'modified' (methylated) bases.
3. hydrolyse DNA molecules foreign to the host cell.
4. recognize short, specific base sequences within the DNA molecule.
5. are used in experiments involving the construction of recombinant DNA molecules.

1. False
They use double-helical DNA as substrate, hydrolysing a phosphodiester bond in each strand.

2. False
Normally, when the DNA of the host cell is 'modified' (e.g. methylated at an A or a C residue), it is protected from cleavage. However, there are exceptions in that a few restriction endonucleases specifically require methylated bases or do not distinguish between methylated and unmethylated bases.

3. True
Bacterial restriction enzymes degrade foreign DNA which is unmodified (see 2).

4. True
Restriction enzymes generally recognize symmetrical sequences of 4—6 base pairs. The nucleotide sequences of the two chains within the specific region are usually identical ('palindromic') when read in the same direction (e.g. $5' \rightarrow 3'$). The specificity of the *E. coli* restriction enzyme EcoRI is as follows:

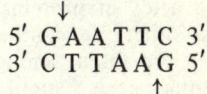

$$\downarrow$$
$$5'\ G\ A\ A\ T\ T\ C\ 3'$$
$$3'\ C\ T\ T\ A\ A\ G\ 5'$$
$$\uparrow$$

5. True
The method is a useful one in that any particular restriction enzyme always cleaves DNA at the same sequence and many produce molecules with identical 'sticky ends'. In this way, for example, viral DNA and foreign DNA, each previously treated with the restriction enzyme, will base-pair at overlapping ends. The strands are then joined by DNA-ligase, thus producing the 'recombinant DNA'.

13.13 During the process of protein biosynthesis in eukaryotic cells:

1. the nucleic acid-containing components required are ribosomes, tRNAs and mRNA.
2. an energy supply in the form of ATP and UTP is required.
3. a supply of NADPH is required.
4. the type of protein synthesized depends only on the type of mRNA present.
5. the absence of even one of the 20 amino acids will prevent the process from proceeding.

1. True
Once the mRNA has been produced in the nucleus, DNA is not involved in the actual process of protein synthesis.

2. False
An energy supply is certainly required, but in the form of GTP. ATP is required in the formation of aminoacyl tRNA, and GTP is required at two stages in the linking of amino acids to form a peptide.

3. False
NADPH is involved in the biosynthesis of many compounds but not proteins.

4. True
For example, a mixture of duck ribosomes, rat tRNAs and baboon mRNA, in the presence of the necessary enzymes, will result in the biosynthesis of baboon-type protein.

5. True
For the vast majority of proteins, the presence of all of the 20 amino acids in proteins is required for protein synthesis, otherwise synthesis will stop at the codons coding for absent amino acids. Proteins with 'missing' amino acids cannot be synthesized. Exceptions would be unrepresentative proteins containing only some of the possible naturally occurring amino acids.

13.14 Protein biosynthesis in prokaryotic cells is inhibited by:

1. tetracycline.
2. cycloheximide.
3. methotrexate.
4. chloramphenicol.
5. theophylline.

1. True
Although tetracycline inhibits protein synthesis by both 70S and 80S ribosomes, most mammalian cells do not accumulate it. In contrast, sensitive bacteria can accumulate it by an active process. It is therefore useful as an antibiotic.

2. False
Cycloheximide inhibits the peptidyl transferase activity located on the 60S subunit of the eukaryotic (80S) ribosome. It does not affect the peptidyl transferase of 70S ribosomes (see 4).

3. False
Methotrexate (amethopterin) has no direct effect on protein biosynthesis. It is a very potent inhibitor of dihydrofolate reductase and prevents the regeneration of tetrahydrofolate. As a consequence, the synthesis of deoxythymidylate (dTMP), and therefore DNA synthesis, is inhibited. Methotrexate is a useful drug in cancer chemotherapy.

Methotrexate

4. True
Chloramphenicol inhibits the peptidyl transferase located on the 50S subunit of prokaryotic ribosomes. Mitochondrial ribosomes are also moderately sensitive to the action of chloramphenicol; however, the compound still has some uses as an antibiotic.

5. False
Theophylline (1,3-dimethylxanthine) inhibits the metabolism of cyclic AMP to AMP, a reaction catalysed by the enzyme cyclic AMP phosphodiesterase. Theophylline and the structurally related compound, caffeine, will therefore potentiate the actions of hormones that use cyclic AMP as a second messenger.

Section 14

COMPLEX LIPIDS

14.1 Which of the following is true of chylomicrons?

1. The range of diameters encountered is 30–500nm.
2. Chylomicrons contain about 50% lipid and 50% protein.
3. Over 80% of the lipid content of chylomicrons constitutes triacylglycerols.
4. They have a density in the range 1.006–1.063g/cm³.
5. They contain small amounts of free cholesterol.

1. True
Chylomicrons carry triacylglycerols from the intestine to other tissues and the range of diameters encountered is indeed 30–500nm.

2. False
Protein represents only about 2% of chylomicrons. The rest is lipid.

3. True
See 1 above.

4. False
Chylomicrons have the lowest density (less than $0.94g/cm^3$) of all the plasma lipoprotein complexes. The range $1.006–1.063g/cm^3$ is that of the so-called LDL or low density lipoproteins.

5. True
A figure of about 2% is typical. In addition, about three times this amount is present as cholesterol esters. Other plasma lipoproteins have proportionately much more cholesterol and cholesterol esters.

14.2 The compound shown below is derived from glycerol.

$$
\begin{array}{l}
\mathrm{CH_2 - OCO - R_1} \\
\quad | \\
\mathrm{R_2 - CO.O - CH} \qquad \mathrm{O} \\
\quad \qquad \qquad | \qquad \quad \| \\
\qquad \qquad \mathrm{CH_2 - O - P - O - R_3} \\
\qquad \qquad \qquad \qquad | \\
\qquad \qquad \qquad \qquad \mathrm{O^-}
\end{array}
$$

Which of the following enzymes would be expected to catalyse the cleavage of this molecule?

1. Pancreatic lipase.
2. Snake venom phospholipase (phospholipase A_2).
3. Plant phospholipase (phospholipase D).
4. Acid phosphatase.
5. Lipophosphodiesterase (phospholipase C).

1. False
Pancreatic lipase hydrolyses tri- and di-glycerides.

2. True
The R_2-containing moiety would be cleaved.

3. True
The R_3-containing moiety would be cleaved, leaving the phosphate group attached to the glycerol.

4. False

5. True
The moiety $R_3 - OPO_3^-$ would be removed.

14.3 A phosphoglyceride such as lecithin is made up from glycerol, phosphate, and:

1. one molecule of long chain fatty acid and choline.
2. two molecules of long chain fatty acid and choline.
3. three molecules of long chain fatty acid and choline.
4. three molecules of long chain fatty acid and ethanolamine.
5. three molecules of long chain fatty acid and inositol.

1. False
Phosphoglycerides contain two fatty acid residues. Phosphoglycerides are derivatives of phosphatidate (diacylglycerol 3-phosphate). The phosphate group of phosphatidate is esterified to the hydroxyl group of one of several alcohols (e.g. serine, ethanolamine, choline, glycerol or inositol).

2. True
The basic structure of a phosphoglyceride is a molecule of glycerol which has two of its hydroxyl groups joined in ester linkages to long chain fatty acid residues. In lecithins, the third hydroxyl group is linked, via phosphate, to choline.

3. False
Only triglycerides (triacylglycerols) contain three molecules of long chain fatty acid.

4. False
See 2 and 3 above.

5. False
See 2 and 3 above.

14.4 A sphingolipid such as sphingomyelin contains the following:

1. glycerol, phosphate and two molecules of long chain fatty acid.
2. glycerol, phosphate, two molecules of long chain fatty acid and choline.
3. sphingosine, phosphate, one molecule of long chain fatty acid and choline.
4. sphingosine, phosphate, two molecules of long chain fatty acid and choline.
5. sphingosine, phosphate, two molecules of long chain fatty acid and inositol.

1. False
Sphingolipids, the second largest group of membrane lipids, are phospholipids. However, they contain the base sphingosine instead of glycerol:

$$H_3C - (CH_2)_{12} - CH = CH - \underset{\underset{OH}{|}}{C} - \underset{\underset{NH_3^+}{|}}{C} - CH_2OH$$

Sphingosine

2. False
See 1 above.

3. True
Sphingolipids consist of sphingosine (see 1), which is attached to choline via a phosphate group and to a long chain fatty acid by an amide linkage.

$$
\begin{array}{l}
NH - OC - R \\
| \\
CH - CHOHCH{=}CH(CH_2)_{12}CH_3 \\
| \\
CH_2 - O \quad O \\
\qquad\quad \backslash \;\; /\!/ \\
\qquad\qquad P \\
\qquad\quad /\;\; \backslash \\
\quad\;\; {}^-O \quad\; OCH_2CH_2N^+(CH_3)_3
\end{array}
$$

The R group is derived from a long chain fatty acid.

4. False
Sphingosine has a long chain resembling that of a fatty acid. Sphingolipids, therefore, like other phospholipids, have two non-polar 'tails'.

5. False
See 3 above. Inositol is a component of phosphatidyl inositol.

14.5 The overall pathway of prostaglandin biosynthesis may be represented as follows:

Membrane phospholipids $\xrightarrow{(A)}$ arachidonic acid $\xrightarrow{(B)}$ prostaglandins

Which of the following is true of this pathway?

1. The enzyme involved at step A is phospholipase A.
2. The reaction, A, is stimulated by anti-inflammatory steroids.
3. The enzyme involved at step B is cyclo-oxygenase.
4. The reaction, B, is stimulated by aspirin.
5. The reaction, A, is the rate-limiting step in prostaglandin synthesis.

1. True
Cholesterol esters containing arachidonic acid may also serve as precursors of arachidonic acid.

2. False
Just the opposite. The hydrolysis of membrane phospholipids by phospholipase A is *inhibited* by anti-inflammatory steroids such as hydrocortisone and betamethasone.

3. True
The key enzyme system of prostaglandin biosynthesis is the prostaglandin synthase complex. The first step is catalysed by the microsomal cyclo-oxygenase component of the synthase and involves cyclization of carbon 8 to carbon 12 of arachidonate.

4. False
Just the opposite. Aspirin and other non-steroidal anti-inflammatory agents (e.g. indomethacin) *inhibit* the cyclo-oxygenase and so block prostaglandin production. Aspirin acetylates the enzyme.

5. True
Agents are known that stimulate step A and hence stimulate prostaglandin production.

14.6 Many 'sphingolipidoses', which are lysosomal storage diseases, are known. In such diseases, a genetic error results in the failure to catabolize a particular lipid or lipid metabolite. An example is Tay-Sachs disease.

Which of the following applies to the enzymes involved in sphingolipid catabolic pathways?

1. The enzymes in question are hydrolases.
2. The pH optimum of each of the enzymes is in the range pH 7—10.
3. The enzymes in question are glycoproteins.
4. Usually a collection of sphingolipids accumulates when there is a genetic error of the type mentioned above.
5. Most of the enzymes occur as isoenzymes.

1. True
Lysosomal enzymes are characteristically hydrolases — examples are β-galactosidase, neuraminidase, hexosaminidase and sulphatase.

2. False
Being lysosomal enzymes, the pH optima of the hydrolases are in the range pH 3.5—5.5.

3. True
In addition to being glycoproteins, many of the enzymes are bound to the lysosomal membrane.

4. False
In the individual metabolic defects, a single catabolic enzyme is missing and usually only a single sphingolipid accumulates.

5. True
Hexosaminidase, for example, occurs in A and B forms.

14.7 The uptake of low-density lipoproteins (LDL) by animal cells:

1. involves interaction with specific, cell-surface receptors.
2. occurs by endocytosis.
3. occurs by facilitated diffusion.
4. involves the action of lipoprotein lipase.
5. provides the primary source of cholesterol for these cells.

1. True
LDL binds to specific receptors located in specialized regions of the plasma membrane ('coated pits'). These regions pinch off to form vesicles which transport the LDL to lysosomes. The receptor proteins are returned to the plasma membrane.

2. True
The general process described in (1) is referred to as 'receptor-mediated endocytosis' and is involved in the uptake of a variety of macromolecules through interaction with specific membrane receptors. The uptake of insulin (and perhaps other peptide hormones) by receptor-mediated endocytosis may play a role in some of the physiological actions of the hormone.

3. False
Uptake is by endocytosis not facilitated diffusion (see 2).

4. False
Lipoprotein lipase provides a mechanism for the uptake (particularly by adipose tissue) of free fatty acids, from circulating triglycerides. The substrate is provided in the form of very low density lipoprotein (VLDL) or chylomicrons. The enzyme is located at the endothelial cell surface and hydrolyses the triglyceride to free fatty acids and glycerol. The free fatty acid can then be re-esterified to triglyceride within the tissue.

5. True
After endocytosis, the LDL is transported to lysosomes and degraded, releasing free cholesterol which can be used for membrane biosynthesis within the cell. Raised, intracellular levels of free cholesterol will inhibit the synthesis of LDL receptors and the synthesis of cholesterol *de novo*. Individuals with an inherited deficiency of LDL receptors are unable to take up LDL from the blood. The resulting high levels of circulating cholesterol may predispose these individuals to premature atherosclerosis.

14.8 Gangliosides:

1. are phospholipids.
2. contain sialic acid (e.g. *N*-acetylneuraminic acid).
3. are typified by sphingomyelin.
4. are present in plasma membranes.
5. are elevated in Tay-Sachs disease.

1. False
Gangliosides are glycolipids. The backbone of the lipid (ceramide) is composed of a long-chain aliphatic amine (sphingosine) to which an acyl chain is attached. The hydrophilic portion of the lipid consists of an oligosaccharide chain containing one or more acidic sugars.

2. True
The acidic sugars in gangliosides are sialic acids.

3. False
Sphingomyelin, like gangliosides, has a backbone of ceramide. However, the hydrophilic portion of sphingomyelin consists of phosphorylcholine rather than sugars. It is therefore a phospholipid.

4. True
They are found in the outer half of the bilayer. Cholera toxin binds to the cell surface by interacting with one particular ganglioside, G_{M1}.

5. True
Tay-Sachs disease is an inherited disorder of ganglioside metabolism. The missing enzyme is lysosomal β-N-acetylhexosaminidase A which removes the terminal N-acetylgalactosamine residue of the ganglioside. The disease is usually fatal within the first years of life.

Section 15

MEMBRANE STRUCTURE AND FUNCTION

15.1 Membrane lipids:

1. do not contain unsaturated fatty acids.
2. are amphipathic.
3. diffuse easily across the lipid bilayer.
4. may be covalently linked to carbohydrate residues.
5. are arranged symmetrically in the membrane bilayer.

1. False
The fatty acid chains in membrane phospholipids and glycolipids can be saturated or unsaturated. The degree of unsaturation of the fatty acid chains affects the 'fluidity' of the membrane.

2. True
The membrane lipids orient themselves with the polar head groups facing the aqueous environment. In phospholipids such as phosphatidyl choline, phosphatidyl serine or phosphatidyl ethanolamine, the polar group is the substituted phosphate. Even cholesterol has a polar head group (the ring hydroxyl).

3. False
Lipid molecules do not move between the two sides of the bilayer at an appreciable rate.

4. True
Carbohydrate-containing lipids (glycolipids) make up about 5% of the lipid molecules in the outer half of the bilayer. They are not present in the cytoplasmic face of the membrane.

5. False
The lipid bilayer is asymmetrical. The glycolipids (see 4) provide a good example of lipid asymmetry.

15.2 Membrane proteins:

1. are generally synthesized on free ribosomes.
2. cannot be removed from the membrane in an active form.
3. can diffuse in the plane of the lipid bilayer.
4. are arranged symmetrically across the lipid bilayer.
5. may comprise 50% or more of the mass of the membrane.

1. False
Integral membrane proteins, like secreted proteins, are generally synthesized on membrane-bound ribosomes of the endoplasmic reticulum.

2. False
Membrane proteins include enzymes, receptors and transport proteins. Some membrane proteins ('peripheral proteins') can readily be removed from the membrane by extraction with salt solutions. Even proteins firmly embedded in the membrane ('integral proteins') can often be removed from the membrane in an active form by treatment with e.g. detergent. In the latter case, the lipid of the membrane is replaced by a micelle of detergent.

3. True
The fluid nature of the membrane allows the proteins

to diffuse rapidly in the plane of the lipid bilayer (but not across it). In this way, membrane proteins can interact with one another.

4. False
For example, adenylate cyclase is predominantly associated with the inner (cytoplasmic) half of the lipid bilayer.

5. True
The proportion of protein can vary from less than 25% to approximately 75% depending on the type of membrane. The mitochondrial inner membrane is about 75% protein.

15.3 Integral membrane proteins:

1. can extend all the way through the lipid bilayer.
2. have a relatively high content of hydrophobic residues.
3. always have lipid covalently attached.
4. are called 'lectins' when they contain covalently bound carbohydrates.
5. do not contain regions of α-helix.

1. True
A portion of each integral membrane protein is embedded in the lipid bilayer. Most, and perhaps all, such proteins span the membrane from one side to the other. Transport and receptor proteins provide typical examples.

2. True
The relatively high hydrophobic content is consistent with the organization of an integral membrane protein within the lipid bilayer, where water is rigorously excluded.

3. False
Lipid is covalently attached to some but by no means all such membrane proteins.

4. False
Lectins are proteins, particularly abundant in plant seeds, that recognize and bind specific sequences of sugar residues. They will therefore interact with cell surface glycoproteins.

5. False
Membrane proteins tend to form α-helical or β-sheet

configurations within the membrane, maximizing H-bonding between peptide bonds. The protein, bacteriorhodopsin, in the purple membrane of *Halobacterium halobium*, crosses the bilayer as seven separate α-helices. A peptide of about 20 amino acids in an α-helical structure is sufficient to cross the bilayer.

15.4 The Na^+/K^+ ATPase:

1. transports 2 Na^+ and 2 K^+ ions for each molecule of ATP hydrolysed.
2. is phosphorylated on an aspartyl residue in the course of the reaction.
3. is inhibited by digitalis glycosides.
4. is inhibited by valinomycin.
5. provides the energy, through ATP hydrolysis, for the active transport of glucose into *Escherichia coli*.

1. False
The ATPase pumps 3 Na^+ ions out for every 2 K^+ ions pumped in. In this way, the ATPase contributes towards maintaining the membrane potential (inside negative relative to outside).

2. True
During hydrolysis of ATP, the terminal phosphate is transferred to an aspartyl residue in the presence of Na^+. Dephosphorylation requires K^+.

3. True
Digitalis glycosides (plant steroids similar to ouabain) inhibit the K^+-dependent dephosphorylation of the enzyme at very low concentrations. As a result of inhibiting the Na^+/K^+ pump, digitalis increases the force of contraction of heart muscle. It has therefore been used in the treatment of congestive heart failure.

4. False
Valinomycin is an example of an *ionophore* which can bind K^+ and transport it across the cell membrane. It does not directly affect the ATPase.

5. False
The active transport of glucose into *Escherichia coli* is directly coupled to the phosphorylation of the sugar to form glucose 6-phosphate. However, phosphoenolpyruvate (and not ATP) is the phosphate donor. The active transport of glucose in intestine and kidney *does* require the Na^+/K^+ ATPase.

15.5 The uptake of glucose by the erythrocyte:

1. is an example of facilitated diffusion.
2. requires hydrolysis of ATP.
3. is saturable.
4. is inhibited by ouabain.
5. is analogous to the uptake of glucose by *Escherichia coli*.

1. True
The accumulation of glucose down its concentration gradient is more rapid than can be explained by simple diffusion.

2. False
Transport of substances by facilitated diffusion does not require energy since it occurs down a concentration gradient. Therefore hydrolysis of ATP does not take place.

3. True
The uptake of glucose shows typical Michaelis-Menten kinetics, implying the involvement of a transport protein in the membrane.

4. False
Ouabain inhibits the Na^+/K^+ ATPase which is only involved in the *active* transport of glucose.

5. False
The phosphotransferase sugar transport system of *Escherichia coli*, involving phosphoenol pyruvate, is an example of *group translocation*.

15.6 The glucose transport system of the epithelial cells of the small intestine:

1. can also transport fructose.
2. is inhibited by 2,4-dinitrophenol.
3. is inhibited by phloridzin.
4. is stimulated by insulin.
5. is dependent on co-transport of Na^+.

1. False
As with enzymes, the gut active transport system for glucose exhibits substrate specificity: glucose and galactose are transported but fructose is not. The inborn error of hexose transport known as 'glucose-galactose maladsorption' therefore does not affect fructose transport.

2. True

The small intestine, like the kidney proximal tubule, accumulates glucose against a concentration gradient. The transport process is therefore energy-dependent and is inhibited by the uncoupling agent, dinitrophenol. Inhibitors of electron transport (e.g. cyanide, oligomycin) will also inhibit the active transport of glucose. This may be contrasted with the passive diffusion of glucose into other tissues.

3. True

The toxic glycoside phloridzin is a competitive inhibitor of active glucose transport. The compound will also prevent sugar reabsorption by the kidney tubule resulting in the urinary excretion of glucose.

4. False

Although insulin stimulates glucose transport into a number of tissues, e.g. adipose and muscle, it does not affect the active transport system of gut and kidney.

5. True

The active transport of glucose is critically dependent upon the asymmetrical distribution of Na^+ and K^+ across the plasma membrane. Sodium ions decrease the K_m of the transport protein for glucose. The ion gradient is restored to normal through the action of the Na^+/K^+ ATPase.

Section 16

HORMONES AND RECEPTORS

16.1 The metabolism of Ca^{2+} is hormonally regulated by:

1. vitamin D.
2. parathyroid hormone.
3. thyroid hormone.
4. calcitonin.
5. calmodulin.

1. True

The active form of the vitamin, 1,25-dihydroxycholecalciferol (calcitriol), controls the absorption of dietary calcium and its deposition in bone. The vitamin is stored in an inactive form (7-dehydro-cholesterol) in skin and is activated by ultraviolet light. Vitamin D deficiency in childhood results in rickets.

2. True
In addition to vitamin D, calcium metabolism is controlled by two peptide hormones, parathyroid hormone and calcitonin.

3. False
The 'C' cells within the thyroid gland secrete calcitonin but thyroid hormone itself does not affect Ca^{2+} metabolism.

4. True
See 2 and 3. Paget's disease, which is characterized by excessive bone resorption resulting in bone fractures and deformities to the skeleton, responds to treatment with calcitonin. The precise defect in this syndrome is unclear.

5. False
The protein, calmodulin, is an intracellular mediator of Ca^{2+} effects. It is not a hormone.

16.2 Insulin:

1. is synthesized in the β-cells of the pancreatic islets.
2. is composed of two polypeptide chains.
3. is secreted in response to a fall in plasma glucose concentration.
4. activates glycogen phosphorylase in liver.
5. is absent in maturity-onset diabetes.

1. True
Glucagon, another pancreatic peptide hormone, is synthesized in the pancreatic islets, but in the α-cells. The rest of the pancreas, the acinal part, synthesizes trypsinogen and other digestive enzyme precursors.

2. True
Although initially synthesized as a *single* polypeptide chain known as proinsulin, the final product is formed by proteolysis of proinsulin to give two chains (A and B) still linked by two disulphide bridges.

3. False
Insulin secretion is stimulated by a *rise* in plasma glucose concentration.

4. False
Insulin secretion stimulates glycogen *synthesis*.

5. False
Maturity-onset diabetics may have normal or above-normal concentrations of plasma insulin. The diabetes is then a consequence of insulin 'resistance' in the target tissues.

16.3 Steroid hormone action:

1. normally involves metabolism of the steroid hormone to an active derivative in the target tissue.
2. involves the second messenger, cyclic AMP.
3. involves selective gene transcription.
4. requires plasma membrane receptors in target cells.
5. involves covalent attachment of the steroid to a receptor.

1. False
An exception is testosterone which must be reduced to 5α-dihydrotestosterone before it can exert its androgenic action.

2. False
Cyclic AMP is not required for steroid hormone action.

3. True
The principal site of action of steroid hormones such as oestradiol and cortisone is in the cell nucleus rather than the plasma membrane. The hormone-receptor complex binds to the chromatin allowing selective transcription of certain genes.

4. False
Steroid hormones can diffuse across the cell membrane and bind to specific intracellular receptors. There is some controversy whether the unoccupied receptors are present in the cytosol.

5. False
Although the association of hormone with receptor is typically of high affinity, it is a non-covalent association.

16.4 Angiotensin II:

1. is derived from angiotensin I by proteolysis.

2. produces thirst when administered centrally to an animal.
3. is formed in the kidney.
4. regulates the biosynthesis and secretion of cortisone.
5. regulates the biosynthesis and secretion of aldosterone.

1. True
A proteolytic cascade reaction is involved in the formation of the octapeptide hormone, angiotensin II. The protease renin is secreted by the kidney in response to decreased blood pressure or $[Na^+]$. Renin acts on plasma angiotensinogen to release a decapeptide, angiotensin I. As angiotensin I passes through the lungs, a second proteolysis step, catalysed by angiotensin converting enzyme (peptidyl dipeptidase), then removes the C-terminal dipeptide of angiotensin I to form angiotensin II. Inhibitors of angiotensin converting enzyme have some application as antihypertensive agents since angiotensin II increases blood pressure.

2. True
The intense thirst produced by administration of angiotensin is one of the most clearcut examples of a specific behavioural response to a chemical.

3. False
See 1.

4. False
Corticotropin (ACTH) is the hormone that regulates cortisone production in the adrenal cortex.

5. True
See 1. Aldosterone assists angiotensin in the restoration of blood pressure and causes sodium retention in the kidney.

16.5 Glucagon:

1. is synthesized in the α-cells of the pancreatic islets.
2. is secreted in response to a fall in plasma glucose concentration.
3. is composed of two polypeptide chains.
4. stimulates cyclic AMP production in liver and adipose tissue.
5. promotes gluconeogenesis in liver.

1. True
Contrast insulin which is synthesized in the β-cells of the pancreatic islets.

2. True
The main action of glucagon is to maintain the blood sugar level by stimulating the breakdown of liver glycogen.

3. False
Glucagon is composed of a single polypeptide chain of 29 amino acids.

4. True
Like many other peptide hormones, glucagon acts through cell-surface receptors to promote cyclic AMP formation. The effect on adipose tissue is to activate mobilizing lipase by phosphorylation, and hence promote lipolysis.

5. True
The actions of glucagon are directed towards preventing hypoglycaemia during fasting.

16.6 Thyroxine:

1. is derived by proteolysis from a precursor protein, thyroglobulin.
2. contains iodine.
3. exerts its hormonal effects by increasing intracellular calcium levels.
4. binds specifically to a cell-surface receptor.
5. selectively affects transcription in target tissues.

1. True
Thyroglobulin, a 19S glycoprotein, is the primary precursor for thyroxine. Modification of tyrosine residues within thyroglobulin results in the formation and release of the thyroid hormones:

Thyroxine (T$_4$)

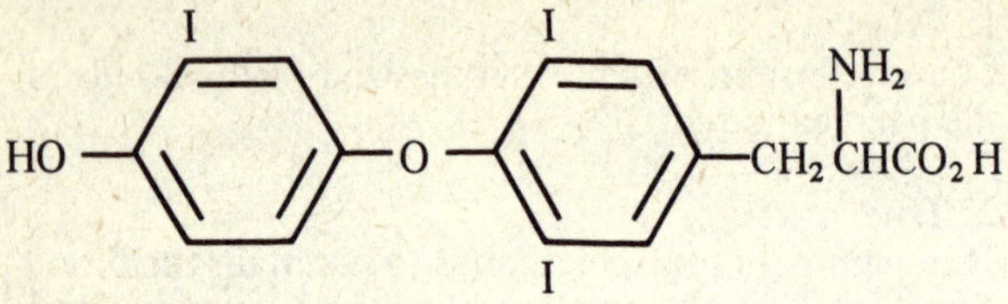

Tri-iodothyronine (T_3)

2. True
See structures above. Iodide is taken up into the thyroid gland, oxidized by a haem-containing peroxidase and transferred to certain of the tyrosyl residues in thyroglobulin. Both T_4 and T_3 are then released by proteolysis: T_3 has the higher physiological activity.

3. False
See 5.

4. False
Receptors for thyroid hormones are intracellular but, unlike steroid hormone receptors, are thought to be located in the nucleus even in the absence of hormone. The thyroid hormones are small, hydrophobic molecules (see 1) and can enter the cell by simple diffusion.

5. True
Activation of the chromatin-bound receptor for thyroid hormone results in an accelerated rate of synthesis of all classes of RNA in target tissues.

16.7 Vasopressin:

1. is structurally similar to oxytocin.
2. is synthesized in the hypothalamus.
3. is stored and released from the adenohypophysis (anterior pituitary).
4. regulates water reabsorption by the kidney.
5. may be absent or defective in diabetes mellitus.

1. True
Vasopressin and oxytocin are both peptides of nine amino acids and differ at only two positions:

Vasopressin   Cys-Tyr-Phe-Glu-Asn-Cys-Pro-Arg-GlyNH$_2$

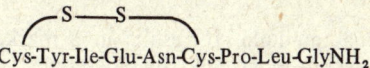

Oxytocin   Cys-Tyr-Ile-Glu-Asn-Cys-Pro-Leu-GlyNH$_2$

Some species, e.g. pig, have a different form of vasopressin in which Arg is replaced by Lys.

2. True
Vasopressin is synthesized in the hypothalamus in the form of a large precursor peptide containing its 'carrier protein', neurophysin 2. Oxytocin is synthesized in a similar manner.

3. False
The hormone precursor is synthesized in the supra-optic nuclei of the hypothalamus and transported along nerve fibres to the neural lobe of the pituitary (*posterior* pituitary). The *anterior* pituitary secretes, among others, β-endorphin, ACTH, luteinizing hormone, follicle stimulating hormone, thyroid stimulating hormone, growth hormone and prolactin.

4. True
Vasopressin is also known as 'antidiuretic hormone'. The antidiuretic effects of vasopressin are mediated by an adenylate cyclase activity located in the collecting ducts of the kidney tubule.

5. False
It is insulin that is absent or defective in *diabetes mellitus*. Diabetes insipidus is a different disease that is characterized by extreme diuresis, e.g. up to 30 l of urine per day. One cause of the disease is an absence of vasopressin, which can be corrected by administration of vasopressin analogues. On the other hand, in nephrogenic diabetes insipidus, vasopressin levels are normal but the target tissues are unresponsive to the hormone. The latter condition cannot be corrected by vasopressin administration and careful control of diet and fluid intake is required.

16.8 A deficiency of thyroid hormone may result from:

1. phaeochromocytoma.
2. Cushing's syndrome.
3. Lesch-Nyhan syndrome.
4. Graves' disease.
5. Hashimoto's disease.

1. False
Phaeochromocytoma is caused by a tumour of the

adrenal medulla. It results in a marked increase in the secretion of the catecholamines which, in turn, causes severe hypertension.

2. False
Cushing's syndrome results from hyperfunction of the adrenal cortex, caused, for example, by a tumour. Some of the effects of excess adrenal steroid hormones include hyperglycaemia, negative nitrogen balance, and retention of sodium and water resulting in oedema. Cushing's *disease* refers to hyperfunction of the adrenal cortex due to increased production of pituitary ACTH.

3. False
Lesch-Nyhan syndrome results from an inherited deficiency of hypoxanthine-guanine phosphoribosyl-transferase, an enzyme of the salvage pathway of purine metabolism. The disease is inherited in a sex-linked recessive manner. The symptoms are dramatic, producing mental retardation and severe self-mutilation. Gout also develops as a consequence of increased turnover of purines and raised uric acid levels.

4. True
Graves' disease is an auto-immune condition characterized by the presence of antibodies that can stimulate the thyroid gland to secrete hormone, resulting in a hyperthyroid state. The antibody duplicates many of the actions of thyrotropin (TSH) but its effects are long-lasting. It has therefore been referred to as 'long-acting thyroid stimulator' (LATS).

5. True
Hashimoto's disease is also an autoimmune disease, in this case resulting in impairment of thyroid function.

16.9 Corticotropin (ACTH):

1. is synthesized in the hypothalamus.
2. is synthesized as a multi-hormone precursor.
3. stimulates formation of cyclic AMP in the adrenal cortex.
4. promotes lipolysis in adipose tissue.
5. stimulates the conversion of cholesterol into pregnenolone.

1. False
ACTH is synthesized in the anterior lobe of the pituitary.

2. True
ACTH is a peptide hormone, 39 amino acids long. It is, however, synthesized as a precursor glycoprotein (pro-opiomelanocortin) of M_r 31 000 which also contains within its sequence other hormones including melanocyte stimulating hormone (MSH), β-lipotropin and β-endorphin. Selective proteolytic processing generates the active hormones.

3. True
Like many other peptide hormones, ACTH acts via cell-surface receptors in its target tissues to activate adenylate cyclase.

4. True
Several hormones, including ACTH, promote lipolysis.

5. True
The conversion of cholesterol into pregnenolone is the rate-limiting step in the synthesis of adrenocortical steroids such as cortisol.

Section 17

IMMUNOLOGY

17.1 The immunoglobulin G molecule (IgG) is composed of two pairs of polypeptides, two 'heavy' and two 'light'. Each 'heavy' chain:

1. has a M_r of about 150 000.
2. consists of four 'domains'.
3. is 'variable' in amino acid sequence for about half its length.
4. is attached to a light chain and to the other heavy chain by disulphide bonds.
5. contains covalently bound carbohydrate.

1. False
The M_r of the whole IgG molecule is about 150 000: the heavy chains have a M_r of about 50 000.

2. True
There are considerable homologies among the domains. This structure may have arisen by gene duplication.

3. False
Antibody specificity for a particular type of antigen

arises because the body is capable of synthesizing many thousands of different types of antibody molecule. Structures can thus be built in which the antibody surface is complementary to that of the antigen. Only the amino-terminal quarter (approximately) of the heavy chains varies from antibody to antibody. The rest of the polypeptide is practically identical in all immunoglobulins of the same class (here IgG). The C-terminal three-quarters of the heavy chain polypeptide is 'constant' in this sense.

4. True
All classes of immunoglobulin polypeptides are linked by disulphide bonds: the number and their disposition along the polypeptide is characteristic of the type of immunoglobulin.

5. True
Covalently bound carbohydrate is attached near the mid-point of each of the heavy chains.

17.2 The immunoglobulin molecule (IgG) is composed of two pairs of polypeptides. Each 'light' chain:

1. has a M_r of about 50 000.
2. will be a λ or a κ type depending only on its origin and not on the type of antigenic specificity.
3. contains two 'domains'.
4. would be 'variable' in amino acid sequence for approximately half its length.
5. would be 'constant' in amino acid sequence for its C-terminal half.

1. False
The M_r is about 25 000: heavy chains have a M_r of about 50 000.

2. True
In all types of immunoglobulin, the light chains are always either λ or κ, which differ in their 'constant' regions.

3. True
One domain, that involved in forming the antigen-binding site, is variable, the other is constant.

4. True
The N-terminal half is 'variable' as it forms half the antigen binding site. Variability from one IgG molecule to another results in the different antigenic specificities.

5. True
The *C*-terminal half does not vary in amino acid sequence with different antigenic specificities.

17.3 Several classes of immunoglobulin molecule are found in the human. Which of the following is true?

1. IgM is found only in the blood.
2. IgA is found only in the blood.
3. IgE is found in association with mast cells in the skin.
4. Only IgG can cross the placental barrier.
5. IgG is the first type of humoral antibody to appear in the serum in response to antigenic challenge.

1. True
Unlike IgA which is found in secretions, IgE which is found in skin, and IgG which crosses the placenta, IgM is found only in serum.

2. False
IgA is present in the blood but is also found in tears, saliva, colostrum and other secretions. It presumably plays a part in 'front-line' defence against invading antigens.

3. True
The majority of IgE is in the skin, but very small amounts are also present in the serum. In certain allergic conditions (e.g. hay fever), serum levels of IgE increase.

4. True
Presumably there are receptors that allow placental transfer to take place. Thus, in the first few days of life, maternal antibody protects the infant until its own immune system becomes operational.

5. False
Almost invariably, IgM is the first type of antibody to be detected in serum.

17.4 Antibody diversity, which enables the body to respond to many different types of antigenic challenge, comes about because:

1. we have one gene for each of the types of antibody molecule we can make.
2. we have a single gene which is susceptible to mutation, producing immunoglobulins with new

amino acid sequences, some of which have a complementary shape to the antigen molecule.

3. we have several genes for the different parts of the 'variable' region of the antibody molecule and, by joining these in various ways, extensive diversity is generated.

4. the IgG molecule is flexible because its -Pro-Pro-Pro- sequence in the hinge regions can adapt its shape to fit an antigen molecule.

5. any IgG molecule only needs to attach to a small region of the antigen molecule, therefore a high degree of specificity is not required.

1. False
In view of the number of antibodies the body can make, it seems unlikely that there would be enough DNA for there to be one gene for each type of antibody.

2. False
Although there may be some 'mutation' this is not the whole explanation. In any case, there are several genes (see 3).

3. True
We have one gene for each of the constant region domains (in heavy and light chains), but many genes for regions of the 'variable' section. In addition, when the genes are spliced, because of slightly different ways of doing this, even more diversity is generated.

4. False
The prolines at the hinge regions do not have this function and indeed their role is unclear. There may be a change of shape (conformation) when antibody binds to antigen, but this is thought to have little to do with the recognition of antigenic site by antibody.

5. False
Although the IgG combines with only a small area on the surface of the antigen molecule, the combination is nonetheless highly specific.

17.5 In the lymphatic system, antibodies for circulation in the serum are produced by:

1. all macrophages.
2. macrophages that have ingested antigen or fragments of antigen.
3. T lymphocytes in the lymph nodes.
4. B lymphocytes that have been transformed into 'plasma cells'.

5. reticulocytes, in the spleen and bone marrow, before they lose their nuclei and become circulating erythrocytes.

1. False
Macrophages do not produce antibodies.

2. False
Macrophages ingest antigens but they do not produce antibodies, although this was formerly thought to be the case.

3. False
Although T lymphocytes may be involved in the antibody response, they do not themselves produce antibody molecules for circulation.

4. True
After interaction with antigen, a B lymphocyte grows in size, divides and synthesizes and secretes antibody. It is then known as a plasma cell.

5. False
Although all blood cells are derived from the same type of stem cell, reticulocytes never produce antibody.

17.6 Completely monospecific antibody, i.e. one consisting of a single type of molecule, may be produced by:

1. a single lymph node in culture.
2. a myeloma.
3. a hybridoma.
4. immunizing with a single, pure antigen.
5. using immunoaffinity chromatography to isolate the antibody.

1. False
Lymph nodes in culture continue to produce antibodies but they produce a collection of types of molecule directed towards a variety of antigenic sites.

2. True
A myeloma is a tumour of the antibody-secreting cells. One cell becomes cancerous and divides repeatedly to form a clone of cells all committed to producing the same type of antibody molecule. Unfortunately, it is rarely known which antigen the antibody is against as the process happens spontaneously.

3. True

'Monoclonal' antibodies are produced by fusing a myeloma-type cell with a spleen cell from an immune animal and growing the resultant 'hybridoma' in culture. A single cell is selected from the culture and is allowed to divide to form a clone of cells all producing exactly the same antibody. (In culture, myeloma cells are 'immortal' but normal spleen cells will not survive very long: the hybridoma is both immortal *and* produces the type of antibody required).

4. False

The antigen has several sites on its surface that are 'recognized' as antigenic by the immune system, which responds by producing several types of antibody molecule. Such antibodies are specific to the same antigen molecule, but bind to different regions on its surface.

5. False

The same considerations as in 4 apply, in that a number of types of antibody molecule will be isolated, each of which will react with a different site on the antigen.

17.7 Which of the following techniques may be used to quantify a protein antigen in a solution of a specific antiserum is available?

1. Double diffusion in agar (Ouchterlony).
2. Single radial immunodiffusion (Mancini).
3. Haemagglutination-inhibition.
4. Rocket immunoelectrophoresis.
5. The complement fixation test.

1. False

Double diffusion will detect an antigen but it is almost impossible to make the method quantitative.

2. True

The area of the circle of precipitate is proportional to the amount of antigen present. Calibration is performed with standard antigen of known concentrations.

3. True

The antigen (e.g. a small protein) is covalently coupled to the surface of a red cell. In the presence of antibody to that particular antigen, the red cells clump together. If free antigen is also present from the solution being assayed, this will combine with some of the antibody molecules and decreased clumping will occur.

4. True
The height of the 'rockets' is proportional to the amount of antigen present. Calibration is performed with antigen of known concentration.

5. True
Antibody-antigen complexes 'consume' complement. If a test for complement is available (e.g. lysis of sensitized sheep red blood cells), then the concentration of antigen can be determined. Antibody is titrated with antigen and the resulting complexes tested for their ability to fix complement: the less complex present, the less complement fixed.

17.8 The complement system of the blood:

1. enables antibody molecules to be transported across the placenta.
2. recognizes antibody-antigen complexes and ensures that they are 'marked' for removal from the circulation.
3. recognizes antibody molecules attached to invading micro-organisms ensuring that such cells will be lysed and destroyed.
4. amplifies the response of antibodies to foreign antigens.
5. recognizes cell-wall material in invading and possibly dangerous organisms (e.g. cell wall polysaccharide of yeasts), setting off a sequence of events that will lead to the invader being inactivated.

1. False
Complement has nothing to do with transport.

2. True
The first component of complement combines only with antibody-antigen complexes, not with free antibody (although it will combine with heat-aggregated IgG). This initial act of recognition marks the antibody-antigen complex for reaction with other components in the complement system, and eventually for attack by macrophages.

3. True
As in 2. The antigen can be soluble or, as in this case, insoluble, being part of a micro-organism. The

terminal components of the complement cascade cause lysis of cells.

4. False
It is not amplification of the immune response. The complement system comes into play after antibody-antigen recognition and is concerned with the inactivation and removal of the antigen.

5. True
The 'alternative pathway' of complement fixation does not require an antibody-antigen complex to trigger it off. It responds to materials such as yeast zymosan which interact with component C3.

17.9 The IgM molecule:

1. has a sedimentation coefficient of 7S.
2. is made up of 10 units, each of which is rather similar to a single IgG molecule.
3. is 'decavalent' in its antigen-combining capacity, i.e. is potentially capable of binding 10 antigen molecules.
4. is found in secretions such as tears and saliva, as well as in serum.
5. can be dissociated into its constituent units by treatment with a sulphydryl reagent such as 2-mercaptoethanol.

1. False
It has a sedimentation coefficient of 19S: IgG has a sedimentation coefficient of 7S.

2. False
It is made up of five such units.

3. True
See 2. Each unit is divalent, therefore there are 10 antigen binding sites altogether.

4. False
Only IgA appears in these secretions as well as in serum.

5. True
The five units are joined by a 'J' (for joining) chain and the links are disulphides. Gentle treatment with thiols produces the five units (plus J chain). Prolonged treatment produces individual polypeptides (the heavy and light chains).

Section 18

NERVE AND MUSCLE

18.1 Myosin:

1. comprises the 'thick' filaments of the myofibril.
2. can catalyse the hydrolysis of ATP.
3. exists predominantly as a triple helical structure.
4. associates with actin.
5. is hydrolysed by trypsin to form tropomyosin.

1. True
The 'thick' filaments of the myofibril mainly contain myosin. The 'thin' filaments contain actin, tropomyosin and troponin.

2. True
Myosin is an 'ATPase', the hydrolysis of ATP providing the energy for muscle contraction to take place. Actin enhances the ATPase activity of myosin.

3. False
Myosin consists of a double-headed globular region joined to a two-stranded α-helical coil. Collagen exists as a triple helical structure.

4. True
The actomyosin complex forms when myosin associates with actin. ATP causes the transitory dissociation of actomyosin to actin and myosin.

5. False
Trypsin hydrolyses myosin into two fragments: light and heavy meromyosins. Heavy meromyosin comprises the globular region of the myosin and contains the ATPase activity. Tropomyosin is a component of the thin filaments of the myofibril.

18.2 Actin:

1. is an abundant protein in many non-muscle cells.
2. contains a 'hinge' region, like an IgG molecule.
3. is processed by chymotrypsin to form α-actinin.
4. is present in intestinal microvilli.
5. is the major component of coated vesicles.

1. True
Actin is not found exclusively in muscle. In many

cell types, actin may comprise 10% or more of the total cell protein.

2. False
Actin monomers (G-actin) can associate to form linear helical filaments (F-actin). *Myosin* contains a 'hinge' region.

3. False
α-Actinin is a separate protein from actin. It binds actin at the 'Z-line' of the myofibril. α-Actinin, like actin, is present in many non-muscle cells.

4. True
Actin comprises much of the core of intestinal and kidney microvilli. Other proteins which cross-link the actin filaments are also present.

5. False
Coated vesicles are involved in endocytosis and the movement of membrane material from one cell compartment to another. However, the major component of these vesicles is not actin but a protein called clathrin.

18.3 The naturally occurring peptide, methionine-enkephalin (Tyr-Gly-Gly-Phe-Met):

1. could be hydrolysed by trypsin.
2. could be hydrolysed by cyanogen bromide.
3. is synthesized by a non-ribosomal mechanism.
4. causes the secretion of corticotropin (ACTH) from the pituitary.
5. is the compound normally binding to the membrane receptors that recognize morphine.

1. False
There are no lysyl or arginyl residues in [Met] enkephalin, and therefore no trypsin-sensitive bonds.

2. False
Cyanogen bromide cleaves peptide chains only on the carboxy-terminal side of methionyl residues. In the process, the methionine is converted to homoserine lactone. Since the methionine residue is at the *C*-terminus, cyanogen bromide would not hydrolyse [Met] enkephalin.

3. False
Many of the small peptide hormones and neuro-peptides appear to be synthesized ribosomally, often

in the form of large precursor proteins containing multiple copies of the biologically active peptides. Preproenkephalin, the precursor for the enkephalins, is a typical example. The tripeptide, glutathione (γ-glutamylcysteinylglycine) is an exception in being synthesized by a non-ribosomal mechanism.

4. False
The secretion of ACTH is principally regulated by a 41-residue peptide (corticotropin-releasing hormone or CRH) released from the hypothalamus.

5. True
The discovery of the enkephalins was a consequence of the search for a naturally occurring brain compound capable of interacting with morphine receptors.

18.4 Patients with McArdle's disease (a deficiency of muscle glycogen phosphorylase):

1. accumulate excessive amounts of glycogen in muscle.
2. inherit the disease in an autosomal recessive manner.
3. rarely survive beyond the age of two years.
4. would show little or no rise in blood sugar levels after administration of adrenaline.
5. would show little or no rise in blood lactate levels after vigorous exercise.

1. True
This is an example of one of the glycogen storage diseases.

2. True
The major glycogen storage diseases, except deficiency of liver phosphorylase kinase, which is X-linked, are inherited in an autosomal recessive manner.

3. False
The inability to mobilize muscle glycogen means that patients have restricted capacity for strenuous exercise. Otherwise, these patients develop normally and show few symptoms. Myoglobin in the urine is sometimes seen as a result of damage to muscle cells during exercise.

4. False
The ability of adrenaline to raise blood sugar levels in such patients shows that liver phosphorylase is

functional. Muscle glycogen cannot contribute to blood glucose since the tissue lacks glucose 6-phosphatase.

5. True
This is a consequence of the inability to mobilize muscle glycogen.

18.5 Compounds which function as neurotransmitters include:

1. enkephalin.
2. carnitine.
3. cysteine.
4. γ-aminobutyrate (GABA).
5. 3,4-dihydroxyphenylalanine (DOPA).

1. True
Two forms of enkephalin are known: [Leu] enkephalin (Tyr-Gly-Gly-Phe-Leu) and [Met] enkephalin (Tyr-Gly-Gly-Phe-Met). First identified in brain, these peptides act at the same membrane receptors to which morphine binds.

2. False
Carnitine is involved in the transfer of acyl derivatives across the mitochondrial inner membrane.

3. False
Some sulphur-containing amino acids, for example taurine but not cysteine, may function as neurotransmitters.

4. True
GABA $(H_3N^+.CH_2.CH_2.CH_2.CO_2^-)$, which is formed by decarboxylation of glutamate, is the major inhibitory neurotransmitter in the brain.

5. False
DOPA is the *precursor* of the catecholamine neurotransmitters and hormones (dopamine, noradrenaline, adrenaline). In the neurological disorder, Parkinson's disease, in which there is a deficiency of brain dopamine, administration of DOPA, which is then converted to dopamine, can partly alleviate the symptoms of the disease.

18.6 Monoamine oxidase:

1. is a flavoprotein.
2. is present in the cytosol.

3. catalyses the oxidation of noradrenaline.
4. catalyses the oxidation of γ-aminobutyrate.
5. is inhibited by the benzodiazepine tranquillizers (e.g. Valium).

1. True
The flavin prosthetic group (FMN) of monoamine
oxidase is covalently attached to the enzyme.
Riboflavin deficiency can result in a decreased
ability of the liver to metabolize monoamines.

2. False
Monoamine oxidase is a mitochondrial enzyme and
is commonly used as a marker for the mitochondrial
outer membrane.

3. True
The reaction catalysed is:

$$RCH_2NH_2 + O_2 + H_2O \rightarrow RCHO + H_2O_2 + NH_3$$

The aldehyde product of the reaction is further
metabolized by oxidation (catalysed by NAD^+-
dependent aldehyde dehydrogenase) or reduction
(catalysed by NADPH-dependent aldehyde reductase).

4. False
The inactivation of γ-aminobutyrate (GABA), an
inhibitory neurotransmitter, is catalysed by GABA
transaminase. Elevation of GABA levels in the
brain, e.g. by inhibition of GABA transaminase,
may provide some protection against epileptic
convulsions. Inhibition of glutamate decarboxylase which synthesizes GABA can provoke seizures
in animals.

5. False
Anti-anxiety drugs such as Valium bind to membrane receptor sites in the brain, potentiating the
inhibitory effects of GABA (see 4). Inhibitors of
monoamine oxidase, on the other hand, have been
used clinically in the treatment of depression.
However, such inhibitors also reduce the ability of
the liver to metabolize dietary amines. Failure to
metabolize the powerful hypertensive agent
tyramine (*p*-hydroxyphenylethylamine), present in
cheese and wine, used to result in several fatalities a
year in patients treated with monoamine oxidase
inhibitors.

**18.7 Available energy for muscle contraction is
obtained during anaerobic glycolysis from the following individual steps:**

1. the formation of fructose 1,6-bisphosphate from fructose 6-phosphate.
2. the conversion of 3-phosphoglyceraldehyde to 1,3-glycerate bisphosphate (1,3-diphosphoglycerate).
3. the conversion of 1,3-diphosphoglycerate to 3-phosphoglycerate.
4. the conversion of phosphoenolpyruvate to pyruvate.
5. the reduction of pyruvate to lactate.

1. False
The formation of fructose 1,6-bisphosphate, catalysed by phosphofructokinase, uses energy and requires ATP.

2. False
The formation of 1,3-diphosphoglycerate from 3-phosphoglyceraldehyde is an oxidation step which requires inorganic phosphate and reduces NAD^+ to NADH. However, under conditions of anaerobic glycolysis, oxidative phosphorylation cannot occur and the NADH is recycled back to NAD^+ by the formation of lactate.

3. True
In the formation of 3-phosphoglycerate from 1,3-diphosphoglycerate, catalysed by phosphoglycerate kinase, the phosphate group passes to ADP. The ATP so formed represents available energy for the cell. Note that two molecules of ATP are formed at this step for each molecule of glucose entering glycolysis.

4. True
The conversion of phosphoenolpyruvate (PEP) to pyruvate, by pyruvate kinase, is the second of the two ATP-producing reactions of glycolysis. Note that in this reaction the 'kinase' is working in the direction of ATP formation rather than ATP utilization, which contrasts with the 'normal' direction of the reaction catalysed by phosphofructokinase. As in 3, two molecules of ATP are formed at this step for each molecule of glucose that is metabolized.

5. False
The formation of lactate from pyruvate during anaerobic glycolysis occurs so that, in the absence of oxygen, the NADH may be reconverted to NAD^+, allowing further glycolysis to take place. This is necessary because the total amount of NAD^+ and NADH in the cell is small and is limited.

18.8 The sarcoplasmic reticulum:

1. is the site of protein synthesis in muscle cells.
2. is present in skeletal and cardiac muscle but not in smooth muscle.
3. initiates the contractile process in muscle by binding Ca^{2+} ions.
4. is the site of ATP production in muscle cells.
5. contains a Ca^{2+}-binding protein, calsequestrin.

1. False
The sarcoplasmic reticulum is a modified form of smooth endoplasmic reticulum and does not contain bound ribosomes.

2. False
All muscle cells contain sarcoplasmic reticulum although there is proportionately less in smooth muscle.

3. False
Quite the opposite: it is the *release* of Ca^{2+} from the sarcoplasmic reticulum into the cytosol (sarcoplasm) that initiates contraction.

4. False
On the contrary, the sarcoplasmic reticulum contains a Ca^{2+}-dependent. ATPase providing the energy for transporting calcium (cf. the Na^+/K^+ ATPase of the plasma membrane). Muscle cells, especially in heart, are generally rich in mitochondria for ATP production, although the quantity of ATP in muscle can sustain contraction for less than a second. Phosphocreatine serves as a 'reserve' of energy in vertebrate muscle.

5. True
Calsequestrin is a highly acidic protein ($M_r = 44\ 000$), present in the lumen of the sarcoplasmic reticulum, that has more than 40 binding sites for Ca^{2+} ions.

18.9 Smooth muscle:

1. is striated in appearance.
2. contains the troponin system.
3. contains tropomyosin.
4. uses Ca^{2+} ions to regulate the contractile process.
5. relaxes in response to an increase in intracellular cyclic AMP.

1. False
Only skeletal and cardiac muscle cells appear striated when observed in the microscope. The normal organization of actin and myosin in smooth muscle does not result in a striated appearance.

2. False
The troponin sytem is present only in striated muscle.

3. True
Tropomyosin, which regulates the interaction of actin and myosin, is present in all types of muscle.

4. True
In smooth muscle, Ca^{2+}, in combination with calmodulin, activates the enzyme 'myosin light-chain kinase'. The resultant phosphorylation of myosin light chain initiates the contractile cycle.

5. True
Stimulation of the β-type of receptors for adrenaline (β-adrenergic receptors), which raises intracellular cyclic AMP levels, causes relaxation of smooth muscle. The effect is probably a result of phosphorylation of myosin light-chain kinase which decreases its ability to activate myosin.

Section 19

NUTRITION

19.1 For nutritional well-being, the mammalian diet must contain, in addition to vitamins, certain specific:

1. amino acids.
2. fatty acids.
3. purines and pyrimidines.
4. monosaccharides.
5. metal ions.

1. True
All mammals require certain essential amino acids in the diet. Adult humans require these essential amino acids in amounts varying from 0.5 to 2g per day, usually ingested in the form of protein. The amino acids usually considered to be essential are lysine, methionine, leucine, isoleucine, valine, phenylalanine,

208

tryptophan, histidine and threonine. In addition, arginine is required by infants and growing children.

2. True
Mammals can readily synthesize oleic acid, a fatty acid containing one unsaturated (double) bond. However, they cannot introduce a second double bond as is found in linoleic and γ-linolenic acid and these compounds must be provided in the diet. They are therefore known as essential fatty acids. Other polyunsaturated fatty acids, α-linolenic and arachidonic acids can be synthesized from dietary linoleic acid. The essential fatty acids are precursors of prostaglandins. The oil of the evening primrose (*Oenothera*) is especially rich in γ-linolenic acid and has been claimed to be beneficial in the treatment of a number of diseases, e.g. multiple sclerosis. Such claims are, at present, unsubstantiated.

3. False
There is no dietary requirement for any specific purine or pyrimidine.

4. False
There is no specific requirement for any particular sugar or derivative in the diet since they are readily formed from glucose (apart from vitamin C which is required in the diet of humans and certain other mammals).

5. True
Many metal elements are required in human nutrition. Those for which there is a major requirement (at least 100mg per day) include calcium, magnesium, potassium and sodium. Others, so-called trace elements, include iron, zinc, manganese and copper. Calcium (ideally about 1g per day for humans) is required for bone formation; iron and copper are needed for haem-protein synthesis and sodium and potassium are required for the maintenance of membrane potential and electrolyte balance. Many enzymes require metal ions as cofactors.

19.2 The approximate, recommended daily requirements for vitamins are as follows:

1. thiamine; 1.5mg.
2. niacin (nicotinic acid); 2mg.
3. cobalamin; 3mg.
4. ascorbic acid; 4.5mg.
5. folic acid; 0.4mg.

1. True
The recommended daily amount of thiamine (B_1) is 1—1.5mg but many people probably consume less. The best sources of thiamine are meat, beans, nuts and whole-grain cereal. Bread made from white flour is notoriously low in thiamine (unless supplemented), as is polished rice.

2. False
The recommended intake of niacin is about 20mg per day, although some can be synthesized from tryptophan. Lean meat, fish, peas and nuts are considered to be the best sources.

3. False
The requirement for cobalamin (vitamin B_{12}) is extremely low at approximately 3μg per day. Because the daily requirement is so low and because it can be stored in the liver over long periods, a true dietary deficiency is rarely encountered.

4. False
The recommended daily requirement for ascorbic acid (vitamin C) varies between 20mg and 60mg per day although it is probable that scurvy can be prevented at a lower level of intake than this. However, some authorities claim that a much higher dietary intake of vitamin C (several grams per day) may be beneficial. Citrus fruits, tomatoes, green vegetables and potatoes are all good sources of vitamin C, but it should be remembered that this vitamin is destroyed by prolonged cooking.

5. True
The recommended dietary intake of folic acid is 400μg. Folic acid deficiency is one of the most common forms of vitamin deficiency, particularly in developing countries and among elderly people. It is found in green vegetables, yeast, meat and liver. Like vitamin C, it is destroyed by cooking.

19.3 Ascorbic acid (vitamin C):

1. is a heat-stable, fat-soluble vitamin.
2. is required as a cofactor for the hydroxylation of certain proteins.
3. if deficient in the diet eventually leads to subcutaneous haemorrhaging, easy bruising, and bleeding and painful gums.

4. is excreted in the urine mainly unchanged if ingested in large amounts.
5. is present in high concentrations in cod-liver oil.

1. False
Ascorbic acid is a water-soluble vitamin which is readily destroyed by heating under mildly alkaline conditions.

2. True
Ascorbic acid acts as a cofactor in a number of enzymic hydroxylation reactions including the hydroxylation of proline residues of collagen.

3. True
These are the symptoms of scurvy. Most animals can synthesize ascorbic acid from glucose, but humans and guinea pigs lack one of the enzymes of the pathway and hence require the vitamin pre-formed in the diet. Severe scurvy is uncommon in adults. Infantile scurvy occurs in babies that are bottle-fed on unsupplemented pasteurized milk or on reconstituted milk powders.

4. True
Large amounts of ascorbic acid cannot be stored in the body and excess intake above that required is excreted in the urine. This brings into question claims that a high dietary intake (of several grams per day) should be taken for optimum health.

5. False
Cod-liver oil contains high concentrations of the fat-soluble vitamins A, D, E and K.

19.4 Iron is:

1. absorbed in the oxidized, ferric Fe(III) form.
2. absorbed very efficiently from the intestine.
3. transported in the blood bound to albumin.
4. stored in the body in the form of ferritin.
5. excreted in large amounts in bile and faeces.

1. False
Iron is absorbed only in the ferrous, Fe(II), form. Ferric, Fe(III), salts are in general highly insoluble: for example, the solubility constant of $Fe(OH)_3$ is about 4×10^{-36}.

2. False
Iron absorption in the small intestine is very slow and inefficient and occurs to different extents depending upon the form in which it is present in the diet. It is estimated that only about 5–10% of ingested iron is actually absorbed. Iron deficiency is one of the most common nutritional disorders in developed countries.

3. False
Iron is carried in blood by a specific iron transport protein, transferrin.

4. True
Iron is stored in the form of ferritin, an iron-protein found in large amounts in spleen, liver and bone marrow. In some instances, when the capacity to store iron as ferritin is exceeded, iron accumulates as haemosiderin granules in mitochondria of some tissues.

5. False
Iron is not *excreted* from the body but is efficiently conserved. A very small amount of iron, resulting from the breakdown of haemoglobin, is lost from the body via bile and faeces. Iron is also lost in sloughed-off cells and in menstrual blood.

19.5 Which of the following diseases may result from a dietary deficiency?

1. Pernicious anaemia.
2. Xerophthalmia.
3. Marasmus.
4. Simple goitre.
5. Addison's disease.

1. False
Pernicious anaemia is a disease where a deficiency in the secretion of a glycoprotein (known as intrinsic factor) by the stomach leads to an impaired absorption of vitamin B_{12}. This in turn leads to a deficiency in erythropoiesis. There are also effects on the central nervous system.

2. True
Xerophthalmia ('dry eyes') and the associated keratomalacia (keratinization of the skin and cornea of the eye) are caused by a lack of vitamin A or its dietary precursor, carotene. It is common in a number of developing countries and is often associated with protein-calorie malnutrition.

3. True
Marasmus is caused by chronic calorie malnutrition in children. It affects children at a critical stage in their development (i.e. immediately after weaning) and is characterized by muscle wasting, weakness and arrested growth. It is usually associated with other vitamin and nutritional deficiencies.

4. True
The thyroid gland accumulates iodine from the blood and uses it in the synthesis of thyroid hormones. If iodine is deficient, the thyroid undergoes a compensatory enlargement. This enlargement, which in extreme cases may result in a thyroid weighing several pounds, is known as simple goitre. It is most common in areas where there is a deficiency of iodine in the soil (it was, at one time, referred to as 'Derbyshire Neck').

5. False
Addison's disease is caused by a deficiency in the secretion of adrenal corticosteroids. This leads to a greatly decreased reabsorption of sodium salts in the kidney and their consequent loss in urine with profound disturbances to body fluid volume and composition. Such patients usually show a craving for salt.

19.6 A deficiency of thiamine (vitamin B_1):

1. causes kwashiorkor.
2. is characterized by a roughening and darkening of the skin, a condition known as pellagra.
3. results in peripheral neuritis, muscular weakness and ultimately death.
4. gives rise to an elevated serum pyruvate level.
5. is frequently associated with chronic alcoholism.

1. False
Kwashiorkor results from chronic protein malnutrition in children. Growth is retarded and the tissues become bloated due to fluid retention (oedema) caused by the low level of plasma proteins.

2. False
Pellagra is caused by a deficiency in nicotinamide (niacinamide). It is prevalent where the diet is low in meat or fish. Such diets, as well as being deficient in nicotinamide, are low in tryptophan from which the vitamin can be synthesized.

3. True

These are the symptoms of beriberi, a disease that became endemic in rice-eating countries with the introduction of 'polished' rice into the diet in the early nineteenth century. The discarded husks, removed during the polishing procedure, contain adequate amounts of thiamine to prevent beriberi.

4. True

Thiamine-deficient animals are unable to oxidize pyruvate normally because thiamine pyrophosphate is a coenzyme for the pyruvate dehydrogenase complex. This impairment of pyruvate metabolism is particularly important in the brain which normally relies upon the aerobic oxidation of glucose.

5. True

Because alcoholics obtain a large fraction of their calorie requirement from alcohol, and consequently do not have a balanced diet, thiamine deficiency is a characteristic of chronic alcoholism. This thiamine deficiency may be responsible for some of the neurological disorders accompanying alcoholism.

19.7 Vitamin D, cholecalciferol:

1. can produce toxic effects if taken in excess.
2. is necessary for the absorption of calcium.
3. activates enzymes necessary for the deposition of hydroxylapatite in bones and teeth.
4. under certain circumstances, can be synthesized from cholesterol.
5. absorption from the small intestine is promoted by bile salts.

1. True

Vitamin D toxicity can be produced in experimental animals and occurs in patients receiving excessive doses of the vitamin. If administered, for example, at ten times the daily requirement, demineralization of bone occurs and fractures may result. Serum calcium and phosphate levels are elevated and calcification of soft tissues, as seen in the formation of renal calculi, may occur.

2. True

Vitamin D_3 (cholecalciferol), although not itself biologically active, is converted first to 25-hydroxycalciferol in the liver and then to 1,25-dihydroxycholecalciferol in the kidney. This is necessary for the absorption of calcium from the upper part of the small intestine. A deficiency of vitamin D, and hence a deficiency in calcium absorption, results in rickets in children and osteomalacia in adults.

3. False
The mechanism of hydroxylapatite deposition is unclear. However, vitamin D is not known to have any direct effect on this process.

4. True
Cholesterol readily forms 7-dehydrocholesterol, the inactive precursor of vitamin D. The 7-dehydrocholesterol is converted to cholecalciferol in the skin by reactions that depend upon exposure to the ultraviolet component of sunlight.

5. True
The bile salts are important in the absorption of all fat-soluble nutrients, including the fat-soluble vitamins A, D, E and K.

19.8 A lack of functional 'intrinsic factor', a glycoprotein secreted by the stomach:

1. is caused by a deficiency of dietary vitamin B_{12}.
2. is the most common cause of vitamin B_{12} deficiency.
3. can be compensated for by the administration of vitamin B_{12}.
4. causes iron-deficiency anaemia.
5. is characterized by the presence of high levels of methylmalonate in blood and urine.

1. False
A lack of intrinsic factor *causes* a deficiency of vitamin B_{12}. The condition is known as pernicious anaemia.

2. True
Because the daily requirement for vitamin B_{12} is so low, a true dietary deficiency is extremely rare and the usual cause of a nutritional deficiency is lack of functional intrinsic factor.

3. True
Pernicious anaemia was, for a period, treated by feeding the patient large amounts of vitamin B_{12} in the form of raw liver. It is now treated by injecting purified B_{12} preparations.

4. False

Iron-deficiency anaemia, in which the number of red cells is normal but the haemoglobin level is low, is caused by a nutritional deficiency of iron.

5. True

Methylmalonic acidaemia (and aciduria) is characteristic of pernicious anaemia because vitamin B_{12} (cobamide) is a coenzyme for methylmalonyl CoA mutase. This enzyme normally converts methylmalonyl CoA to succinyl CoA during the metabolism of propionic acid and the amino acids methionine, valine and isoleucine. Methylmalonate accumulation also occurs in individuals having a defective mutase.

19.9 Ethanol:

1. is absorbed mainly by the stomach.
2. can readily be converted to glycogen by the body.
3. may provide over half the calorie intake of heavy drinkers and alcoholics.
4. is oxidized to acetaldehyde as the first step in its metabolism.
5. when taken in excess, inhibits the absorption of thiamine.

1. False

Only about 20% of ethanol is absorbed by the stomach with the rest being absorbed by the small intestine. The rate and site of absorption are influenced by the amount and nature of food present in the stomach. Alcohol is absorbed very rapidly into the bloodstream, with a single dose being detectable after about five minutes and a maximum concentration occurring in half an hour to two hours.

2. False

Alcohol cannot give rise to glucose and therefore cannot be converted to glycogen. Furthermore, alcohol *inhibits* gluconeogenesis. It is metabolized via acetate (and acetyl CoA) which is readily converted to fat and which is highly ketogenic.

3. True

A high consumption of alcohol is frequently associated with a low intake of food. Ethanol has a high energy content which is between that of carbohydrate and that of fat.

4. True

Alcohol is first oxidized to acetaldehyde in the liver

by alcohol dehydrogenase and the acetaldehyde is then further oxidized to acetate by aldehyde dehydrogenase. Both enzymes use NAD^+ as the cofactor. Disulfiram (Antabuse), an inhibitor of aldehyde dehydrogenase, has been used in the treatment of alcoholism. The drug causes an accumulation of acetaldehyde, resulting in severe nausea, after alcohol consumption.

5. False
Alcohol does not normally inhibit the uptake of thiamine. Chronic alcoholics often suffer from thiamine deficiency and, to a lesser extent, from other vitamin deficiencies. This is mainly because they tend to be eating insufficient amounts of food of good nutritional quality.

Section 20

MOLECULAR BASIS OF DISEASE

20.1 An inborn error of metabolism is known in which the enzyme glucose 6-phosphatase is deficient. In patients with this disease, under fasting conditions, the enzyme deficiency would be consistent with:

1. failure of adrenaline to raise blood glucose levels.
2. elevated concentrations of free fatty acids in the plasma.
3. metabolic acidosis.
4. excessive glycogen deposition in the liver.
5. elevated levels of plasma insulin.

1. True
Although adrenaline will stimulate glycogenolysis in liver, no free glucose can be formed in the absence of glucose 6-phosphatase.

2. True
The hypoglycaemic state characteristic of glucose 6-phosphatase deficiency will result in the mobilization of lipids from adipose tissue.

3. True
In this condition, the levels of blood lactate and ketone bodies (acetoacetate, β-hydroxybutyrate) will increase, the net result being a fall in blood pH.

4. True
Glucose 6-phosphatase deficiency (von Gierke's disease) is one of the classical glycogen storage diseases. The inability to mobilize glycogen leads to its accumulation in the liver since glycogen synthetase is activated by glucose 6-phosphate.

5. False
Consistent with the hypoglycaemic state, insulin levels will be below normal.

20.2 Abnormal storage of glycogen results from deficiencies in:

1. phosphorylase.
2. lactate dehydrogenase.
3. phosphofructokinase.
4. amylo-1,6-glucosidase.
5. glycogen synthase.

1. True
A deficiency in either muscle phosphorylase (McArdle's disease) or liver phosphorylase (Hers' disease) will lead to an increased amount of glycogen in the appropriate tissue or organ due to poor mobilization. This is accompanied by the early onset of muscle fatigue in the former case and by fasting hypoglycaemia in the latter.

2 and 3. False
Significant deficiencies in the activities of lactate dehydrogenase and phosphofructokinase, both of which are important glycolytic enzymes, would be fatal.

4. True
A deficiency in amylo-1,6-glucosidase, which is associated with the debranching of glycogen, is known as Cori's disease and occurs in muscle and liver. Abnormal glycogen with short outer branches accumulates and the symptoms are muscle fatigue and hypoglycaemia.

5. True
A lack of glycogen synthase in liver, a known hereditary abnormality, leads to an absence of stored glycogen, acute fasting hypoglycaemia (where the blood glucose level may fall to some 10–15% of normal), and irreversible brain damage if not treated immediately. Strictly speaking, it does not qualify as

one of the glycogen storage diseases because the latter are defined as involving the *accumulation* either of large amounts of glycogen or of glycogen of abnormal structure.

20.3 Galactosaemia:

1. is characterized by an accumulation of galactose and galactose 1-phosphate in blood and other tissues.
2. most commonly results from a deficiency in galactokinase.
3. is associated with mental deficiency and can result in impaired vision.
4. is a genetically linked disease.
5. is less critical in adults than in children.

1. True
Galactose is also excreted in the urine.

2. False
The most common cause of galactosaemia is a deficiency in the enzyme UDP-glucose:galactose 1-phosphate uridylyl transferase (or simply uridyl transferase) which forms UDP-galactose and glucose 1-phosphate. A milder form of galactosaemia can result from a deficiency in either galactokinase or UDP-glucose 4-epimerase (the enzyme converting UDP-galactose to UDP-glucose).

3. True
The enzyme deficiency is critical in infants because they are dependent upon lactose in the milk. If not alleviated by omitting milk from the diet, there is mental retardation and enlargement of the liver. Cataract formation may occur later in life.

4. True
Galactosaemia is an inherited metabolic disease (an inborn error of metabolism). The heterozygote carriers can be detected since they have a lowered (50%) activity of uridyl transferase in liver and, as more conveniently assayed, in white blood cells.

5. True
The disease appears to be less critical in adults. It has been suggested that this is because an alternative route for galactose metabolism develops, although this has been questioned. In any case, the adult is far less dependent on lactose as a source of carbohydrate.

20.4 The urinary excretion of large amounts of phenylalanine and phenylpyruvate, occurring in phenylketonuria, results from:

1. a deficiency in tyrosine transaminase (amino-transferase).
2. a deficiency in phenylalanine hydroxylase.
3. a genetic defect.
4. pyridoxal phosphate deficiency.
5. protein malnutrition.

1. False
A deficiency of tyrosine transaminase is not one of the recognized disturbances of tyrosine metabolism. In phenylketonuria, the phenylpyruvate arises by the transamination of phenylalanine which is probably catalysed by tyrosine transaminase.

2. True
Phenylketonuria is caused by a deficiency in the enzyme hydroxylating phenylalanine to tyrosine. Untreated individuals are almost always severely mentally retarded by the age of one year. Treatment depends upon early diagnosis and the restriction of phenylalanine in the diet to the minimum amount required for protein synthesis and normal development.

3. True
Approximately one person in 70 in the population carries a defective gene for phenylalanine hydroxylase. Therefore, approximately one in 5000 children are born to parents each of whom is a carrier. One in four of these children (i.e. approximately one in every 20 000 births) will be doubly recessive for the defective gene and will therefore have phenylketonuria.

4. False
Pyridoxal phosphate is not a cofactor for phenylalanine hydroxylase.

5. False
Phenylketonuria occurs in individuals on normal diets and is actually ameliorated by a low protein (low phenylalanine) intake.

20.5 A positive test for reducing sugar in the urine:

1. is characteristic of diabetes mellitus.

2. is characteristic of lactose intolerance associated with impaired digestion of lactose.
3. may be due to the presence of certain pentoses.
4. is characteristic of fructosuria caused by a deficiency of fructokinase.
5. may be due to galactose and indicate galactosaemia.

1. True
A test for the excretion of glucose in the urine is traditionally one of the first tests in the diagnosis of diabetes mellitus. The test is now normally done with an indicator system (dipstick or indicator paper) which incorporates the enzyme glucose oxidase and is thus highly specific for glucose.

2. False
Lactose intolerance, caused by a deficiency of intestinal lactase, results in the non-absorption of lactose or its hydrolysis products, heavy bacterial contamination of faeces, and diarrhoea.

3. True
Pentoses ingested in fruit or fruit juices may exceed the low renal threshold for these sugars and be excreted unchanged in the urine. A harmless, genetic idiopathic pentosuria occurs in Ashkenazi Jews.

4. True
In cases of hereditary fructosuria, extremely large amounts of fructose may be excreted in the urine. Before the introduction of specific tests for glucose, this often led to a false diagnosis of diabetes (see 1).

5. True
Galactose is a reducing sugar and is excreted in the urine in cases of galactosaemia. However, other symptoms such as the poor development and general distress of the infant are likely to be the first pointers to galactosaemia.

20.6 Acute hepatitis is normally associated with:

1. a decreased serum GOT/GPT ratio (i.e. glutamate oxaloacetate transaminase/glutamate pyruvate transaminase ratio).
2. a greatly increased serum level of sorbitol dehydrogenase.
3. an increased serum creatine phosphokinase (creatine kinase) activity.
4. an increased serum lactate dehydrogenase activity.
5. an increased serum aldolase activity.

1. True

The GOT/GPT ratio (the DeRitis quotient) is usually about 1.3 in normal serum but may fall to 0.6 or 0.7 in the first two weeks of jaundice. This is because, although there is a 15 to 30-fold increase in GOT, there is a larger increase (25 to 30-fold) in GPT.

2. True

A large increase in the serum level of sorbitol dehydrogenase accompanies liver damage.

3. False

The activity of creatine phosphokinase (CPK) in liver is extremely low compared with that in skeletal muscle, heart and even brain. There is no significant increase in the serum level of CPK after liver damage.

4. True

There is usually a large increase in lactate dehydrogenase (LDH) in the serum when there is liver damage. The LDH is of a different isoenzyme form from that released following heart damage. The two forms, as well as being electrophoretically and chromatographically separable, have different heat stabilities.

5. True

There is an increase, sometimes very large, in the level of serum aldolase. However, this is not particularly diagnostic of liver damage since the enzyme is present in many other tissues.

20.7 A myocardial infarction will lead to an increase in serum levels of:

1. lactate dehydrogenase.
2. creatine phosphokinase (creatine kinase).
3. glucose 6-phosphatase.
4. glutamate-oxaloacetate transaminase (GOT).
5. glutamate-pyruvate transaminase (GPT).

1. True

There is usually a steady increase in LDH activity in the serum over the first two or three days following infarction and a return to normal in eight to ten days, depending upon the severity of the attack. The average increase after infarction is two to five times the upper limit of normal. The LDH that is released into the serum is predominantly the H_4 isoenzyme.

2. True

Following infarction, the CPK level is usually from

three to thirty times the upper limit of normal. In the absence of other muscle damage, an elevated CPK level is strongly indicative of heart damage.

3. False
There is no increase in serum glucose 6-phosphatase activity following damage to heart tissue, nor would one be expected since the enzyme is of very low activity in muscle.

4 and 5. True
Both GOT and GPT activities rise after heart infarction. However, what is more significant is the ratio of these two activities (the DeRitis quotient). The GOT/GPT ratio is normally about 1.3 but is usually well over 2 following an infarction because more GOT is released than GPT.

20.8 An elevated serum alkaline phosphatase level is associated with:

1. Paget's disease.
2. muscular dystrophy.
3. bone tumours.
4. rickets.
5. carcinoma of the prostate.

1. True
A rise in serum alkaline phosphatase level is one of the few biochemical changes observed with Paget's disease, a localized bone disease in which there is an abnormally high rate of bone resorption.

2. False
An increased alkaline phosphatase activity is not associated with muscular dystrophy.

3. True
The presence of bone tumours often leads to very high serum alkaline phosphatase activities, especially in osteosarcoma where the increase may be 20 to 40 times the normal.

4. True
There is usually a rise in serum alkaline phosphatase which occurs before rickets becomes clinically manifest. The level falls following the administration of vitamin D.

5. True
Carcinoma of the prostate does give rise to an elevated

alkaline phosphatase level in serum. However, the activity of acid phosphatase is also raised and this is diagnostically useful in distinguishing between prostatic carcinoma and bone diseases.

20.9 The human haemoglobinopathy, thalassaemia:

1. is another name for sickle-cell disease.
2. results from overproduction of normal haemo-globin (HbA) leading to haemolytic anaemia.
3. results from partial or complete failure to produce one or other of the polypeptide chains of haemo-globin leading to haemolytic anaemia.
4. can result from altered haemoglobin metabolism following an attack of malaria.
5. results from a deficiency of one of the enzymes involved in the biosynthesis of the haem moiety of haemoglobin.

1. False
The term thalassaemia encompasses a whole series of inherited diseases in which various mutations have taken place. In some α-thalassaemias, small amounts of abnormally long α-globin molecules are produced as a result of mutation of a stop codon (UAA) to any one of four sense-codons. In some β-thalassae-mias, no globin mRNA can be detected, whereas in another the β-globin mRNA is present but is not translated. Sickle-cell disease refers to a specific disease in which a point mutation has occurred in the β-chain gene so that amino acid No.6 is valine rather than glutamate leading to altered properties of the haemoglobin (HbS).

2. False
Thalassaemia results from *under*-production of one of the polypeptide chains of haemoglobin.

3. True
Thalassaemia results from an imbalance in the stoi-chiometry of globin synthesis. One type of chain is either absent or very much reduced in amount. The remaining chain accumulates, precipitates and damages the red cell membrane leading to haemolytic anaemia.

4. False
Thalassaemias are inherited diseases. Indeed, sufferers from thalassaemia, like those with sickle-cell disease, are somewhat more resistant to malaria than normal individuals.

5. False
As far as is known, the enzymes for haem biosynthesis are normal: it is the biosynthesis of globin that is abnormal.

Section 21

MISCELLANEOUS

21.1 Carboxylation reactions (CO_2-fixation reactions) in animal cells:

1. require thiamine pyrophosphate.
2. frequently require biotin.
3. usually require energy in the form of a nucleoside triphosphate (e.g. ATP).
4. involve lipoic acid.
5. are often unidirectional and therefore may be important control points in the pathways to which they belong.

1. False
Thiamine pyrophosphate, the coenzyme form of vitamin B_1, functions in several enzymic reactions involving the transient formation of a complex containing the coenzyme and an aldehyde derivative of the substrate. A number of these reactions, such as the pyruvate decarboxylase reaction (in yeast) and the pyruvate dehydrogenase and α-oxoglutarate dehydrogenase reactions of the TCA cycle, involve decarboxylation.

2. True
Biotin is the coenzyme in a number of important carboxylation reactions. The carbon dioxide is carried as a carboxyl group attached to a nitrogen atom of the coenzyme. The biotin molecule is covalently attached to the enzyme via a lysyl residue at the active site. Biotinyl enzymes also require magnesium ions.

3. True
Energy is required to form carbon-carbon bonds. It is provided by the hydrolysis of ATP in the pyruvate carboxylase and acetyl CoA carboxylase reactions.

4. False
Lipoic acid, an essential vitamin or growth factor for

a number of micro-organisms but readily synthesized by higher organisms, is a component of the pyruvate dehydrogenase and α-oxoglutarate dehydrogenase complexes, both of which catalyse *decarboxylation* reactions.

5. True
Two important carboxylation reactions are those catalysed by acetyl CoA carboxylase and pyruvate carboxylase. Each reaction is physiologically uni-directional and catalyses the first step in a biosynthetic pathway. Acetyl CoA carboxylase, catalysing the formation of malonyl CoA for the synthesis of fatty acids, is activated by citrate. Pyruvate carboxylase, catalysing the formation of oxaloacetate either for entry into the TCA cycle or as the first step in gluconeogenesis, is activated by acetyl CoA.

21.2 Penicillin G, from *Penicillium chrysogenum*:

1. is a polypeptide antibiotic.
2. specifically inhibits protein synthesis by bacterial (70S) ribosomes.
3. specifically inhibits the transpeptidase reaction in bacterial cell-wall synthesis.
4. is effective as an antibiotic only against cells that are actively dividing.
5. is destroyed by penicillinase which confers resistance on a range of bacteria.

1. False
Penicillin G consists of a thiazolidine ring fused to a β-lactam ring to which is attached a benzyl group. This benzyl group may be replaced, chemically, by different groups thereby improving the effectiveness of the antibiotic.

2. False
Penicillin has no effect on protein synthesis. Chloramphenicol is an antibiotic that specifically inhibits the peptidyl transferase reaction of protein synthesis in 70S ribosomes.

3. True
Penicillin inhibits the transpeptidase reaction of bacterial cell-wall synthesis by covalently linking to a serine at the active site of the enzyme. Penicillin resembles acyl-D-Ala-D-Ala, one of the substrates of the enzyme.

4. True
Because of its mechanism of action referred to in 3, above, penicillin acts on growing, rather than non-growing cells, i.e. those synthesizing new cell walls.

5. True
A number of bacteria produce penicillinase which renders penicillin ineffective by converting it to penicilloic acid. The gene for penicillinase is often carried on a plasmid and can be transferred between different bacterial species. Penicillins in which the benzyl group has been replaced are usually poor substrates for penicillinase and therefore more effective with organisms that have acquired resistance. A good example is methicillin which contains a dimethoxyphenyl group instead of the benzyl group.

21.3 The pigment, melanin:

1. is formed from tryptophan.
2. is formed from tyrosine.
3. is not formed in albinism.
4. is present in skin in order to absorb UV radiation to protect underlying structures.
5. always contains sulphur derived from cysteine.

1. False
Melanin is not formed from tryptophan (see 2), although at one time this was thought to be the case.

2. True
Tyrosine is oxidized by the enzyme tyrosinase to form first DOPA and then dopaquinone. Polymerization of dopaquinone produces melanin. Tyrosinase is distinct from tyrosine hydroxylase which converts tyrosine to DOPA (but not dopaquinone) in the biosynthesis of the catecholamines (dopamine, noradrenaline, adrenaline).

3. True
In one type of albinism there is no tyrosinase and therefore no melanin. The condition, which is recessive, was first described in 1698 in American Indians.

4. True
In humans this is the case. However, melanin has various functions in other animals, e.g. as camouflage, as a sex-attractant, etc.

5. False
Two forms of melanin are known. The black/brown *eumelanin* is formed by polymerization of dopaquinone alone. The yellow/reddish brown *phaeomelanin* is produced when dopaquinone reacts with cysteine before polymerization occurs.

21.4 The coenzyme pyridoxal phosphate:

1. is a derivative of niacin (nicotinic acid).
2. can be synthesized from simple carbon, nitrogen and phosphate compounds by most bacteria.
3. is a coenzyme for phosphorylation-dephosphorylation reactions.
4. is a coenzyme for transaminase (aminotransferase) reactions.
5. is a coenzyme for many amino acid decarboxylation reactions.

1. False
Pyridoxal phosphate is a derivative of pyridoxine, vitamin B_6.

2. True
Most bacteria can synthesize their own pyridoxine which is, therefore, a growth requirement for relatively few species. The same applies to vitamins.

3. False
Phosphorylations usually involve kinases and ATP; dephosphorylation reactions are normally catalysed by phosphatases. Neither of these reactions involves pyridoxal phosphate.

4. True
The participation of pyridoxal phosphate in transamination reactions probably represents, quantitatively, the major role of this coenzyme.

5. True
Most amino acid decarboxylation reactions require pyridoxal phosphate. Examples are the formation of γ-aminobutyrate (GABA) from glutamate in brain, the formation of histamine from histidine and a number of bacterial amino acid decarboxylase reactions.

21.5 Both haemoglobin and cytochrome *c*:

1. consist of protein, and an iron-containing prosthetic group.
2. carry oxygen.
3. contain iron which is oxidized in each case to the ferric state, Fe(III), during normal physiological function.
4. are characterized by a strong absorption peak close to 410nm.
5. have similar relative molecular masses (M_r values).

1. True
Both haemoglobin and cytochrome c are proteins which have iron present in the form of an iron-porphyrin (haem) prosthetic group. The haem of cytochromes b_1, c_1 and c is the same as that in myoglobin and haemoglobin; that in cytochromes a and b differs slightly.

2. False
Only haemoglobin binds oxygen. Each molecule of haemoglobin co-operatively binds four molecules of oxygen, one to each of the iron atoms of the four haem molecules. Mammalian blood is able to bind up to 20% of its volume of gaseous oxygen. At the lungs, the haemoglobin is almost wholly in the oxygenated form. Haemoglobin leaving the tissues will have released about one-third of its oxygen.

3. False
The iron of functional haemoglobin is always in the ferrous (reduced) form. Oxidation to the ferric state gives rise to methaemoglobin which is unable to carry oxygen. This is in contrast to the iron present in cytochromes which undergoes alternate oxidation-reduction during the electron transport process.

4. True
Because of the extensive conjugation of the unsaturated double bond system, porphyrins have striking absorption bands in both the ultraviolet and visible regions of the spectrum. The absorption maximum near 410nm is characteristic of all porphyrins and is known as the Soret band.

5. False
Haemoglobin is tetrameric with four subunits (two α and two β), each with a relative molecular mass of approximately 16 000. Cytochrome c consists of a single polypeptide chain of about 100 residues and has only one haem group. The relative molecular mass of mammalian haemoglobin is 64 500 and that of cytochrome c is 12 400.

21.6 Flavin adenine dinucleotide (FAD):

1. is a coenzyme for hydrogen transfer reactions.
2. contains two vitamins, riboflavin (B_2) and nicotin-amide.
3. shows an increase in absorption at 340nm when reduced.
4. is bright yellow in colour when in the oxidized form.
5. is a coenzyme for succinate dehydrogenase to which it is tightly bound.

1. True
FAD is a hydrogen acceptor primarily for oxidations of the type $-CH_2-CH_2-$ to $-CH=CH-$ rather than for oxidations consisting of $-CH.OH-$ to $-CO-$. It is also a coenzyme for L-amino acid oxidase which catalyses the oxidation of L-amino acids to the appropriate oxo-acids with the release of ammonia.

2. False
FAD is a dinucleotide because it consists of two base-sugar-phosphate moieties attached to each other by the phosphate groups. One of the nucleotide moieties contains riboflavin (an isoalloxazine derivative attached to the sugar alcohol, ribitol). The other, however, contains adenine as the base, not nicotinamide.

3. False
The increase in absorption at 340nm upon reduction is characteristic of the nicotinamide coenzymes (NAD^+ and $NADP^+$). It is due to the reduction of the nicotinamide part of the molecule resulting from hydrogen transfer. FAD does not contain nicotin-amide.

4. True
FAD is yellow when oxidized and FAD-containing enzymes are also characteristically yellow with an absorption peak at approximately 450nm. This peak is virtually abolished on reduction of FAD and solutions of $FADH_2$ are almost colourless.

5. True
FAD is the tightly bound coenzyme for the mito-chondrial enzyme, succinate dehydrogenase, which catalyses the formation of fumarate from succinate.

21.7 Proteins are commonly purified by:

1. ion-exchange chromatography.
2. paper chromatography.
3. gel filtration (gel exclusion) chromatography.
4. salt precipitation.
5. extraction with phenol.

1. True
Ion-exchange chromatography is used in the purification of most proteins. The most commonly used materials are diethylaminoethyl (DEAE)-cellulose and carboxymethyl (CM)-cellulose. The former is positively charged and used for proteins at pH values above their isoelectric points. Elution is then normally achieved by gradually raising the ionic strength or decreasing the pH of the eluting buffer. CM-cellulose is used with proteins at pH values below their isoelectric points and elution is obtained by raising the pH of the eluting system.

2. False
Paper chromatography is not a recognized step in purifying proteins. However, quite large peptides can often usefully be separated from each other using this technique.

3. True
Gel filtration (gel exclusion) chromatography finds two types of application in the purification of proteins. A gel with a low exclusion limit (such as Sephadex G-25®) often takes the place of dialysis since large molecules (such as proteins) pass quickly through the column and smaller ones (salts and co-enzymes) are retarded. By using a gel with a much higher exclusion limit (Sephadex G-100® or Sephadex G-200®) proteins will move down the column at different rates and will separate from each other in a manner that is determined by size and shape.

4. True
Selective precipitation of proteins by gradually increasing the ionic strength of the solution is one of the earliest and still most frequently used methods in protein purification. Ammonium sulphate is usually the salt that is chosen, mainly because of its high solubility at low temperatures ($515g/l$ at $0°C$).

5. False
Phenol is often used in the purification of DNA which is precipitated at a phenol-water (or ethanol-water) 'interface'. Phenol is not used for the purification of proteins.

21.8 The red cells of blood have a finite lifetime, and the degradation of haem in humans:

1. is incomplete in cases of porphyria leading to the accumulation of uroporphyrinogen I or porphobilinogen.
2. gives rise to bile salts.
3. gives rise to bile pigments.
4. involves the spleen and liver.
5. occurs as a result of the turnover of erythrocytes.

1. False
It is genetic defects in the *biosynthesis* of haem that cause the accumulation of uroporphyrinogen I or porphobilinogen in porphyria.

2. False
Bile salts (glycocholic and taurocholic acids) are derivatives of cholesterol and are neither structurally nor metabolically related to haem.

3. True
Iron is removed from haem and the remaining structure is cleaved to give a linear tetrapyrrole, biliverdin, which is then converted to bilirubin. Both bilirubin and biliverdin are bile pigments. Biliverdin is also responsible for the early blue-green colour of bruises whereas bilirubin is the pigment giving rise to the later yellow colour and also causes the yellowing of skin, eyeballs and urine in jaundice.

4. True
Red blood cells are phagocytosed by macrophages mainly in the spleen, liver and bone marrow and the haem is degraded to bilirubin. Extrahepatic bilirubin is transferred to the liver bound to plasma albumin. In the liver, it is conjugated with glucuronic acid to increase its solubility prior to excretion.

5. True
The normal life of a red blood cell, and therefore of haemoglobin and haem, is approximately 120 days.

21.9 Pancreatic juice:

1. has a pH of approximately 7 to 8.
2. contains bile salts.
3. contains insulin.
4. contains trypsinogen.
5. contains α-1,4-glucosidase (acid maltase).

1. True

The pH of pancreatic juice is normally between 7 and 8. The alkalinity, primarily due to bicarbonate, neutralizes the acidic contents of the digestive tract as they leave the stomach. An adult human secretes about 650ml of pancreatic juice in 24 hours.

2. False

Bile salts (mainly glycocholate and taurocholate) are secreted by the liver and pass, in bile, to the gall bladder. In humans, bile enters the duodenum down the same duct that is used for pancreatic juice. Bile, like pancreatic juice, has a slightly alkaline pH.

3. False

Insulin is secreted by the β-cells of the Islets of Langerhans (the β-islet cells) and passes into the bloodstream, not into the digestive pancreatic juice.

4. True

Pancreatic juice contains trypsinogen, the inactive zymogen precursor of trypsin, and also other zymogens such as chymotrypsinogen, proelastase, procarboxypeptidase and prophospholipase. Trypsinogen, in the first instance, is converted to trypsin by enterokinase, a specific proteolytic enzyme secreted by the small intestine. Once some trypsin is formed it can catalyse the conversion of more trypsinogen to trypsin as well as activate the other zymogens.

5. False

Pancreatic juice contains α-amylase, which, although it hydrolyses α-1,4 linkages of starch and glycogen, is not the same enzyme as α-1,4-glucosidase. The latter is a lysosomal enzyme, a deficiency of which causes Pompe's (generalized) glycogen storage disease.

21.10 In the process of porphyrin biosynthesis:

1. four histidine molecules combine to form a tetrapyrrole.
2. the first identifiable step is the enzyme-catalysed combination of glycine and succinyl CoA to form δ-aminolaevulinic acid.
3. four molecules of δ-aminolaevulinic acid combine to form porphobilinogen.
4. pyridoxal phosphate is required as a cofactor.
5. the same initial steps of the pathway are followed whether the porphyrin is incorporated into haemoglobin, cytochromes, the chlorphylls or vitamin B_{12}.

1. False

Histidine is not involved in porphyrin biosynthesis. The histidine molecule contains an imidazole ring, not a pyrrole ring.

2. True

The first reaction is catalysed by δ-aminolaevulinic acid synthetase with the release of CoA and CO_2 (also see 4):

$$NH_2 CHCO_2 H + HO_2 CCH_2 CH_2 COCoA \longrightarrow$$
$$NH_2 CHCOCH_2 CH_2 CO_2 H + CO_2 + CoA$$

3. False

Two, not four, molecules of δ-aminolaevulinic acid react to form porphobilinogen, four of which eventually combine to form a tetrapyrrole.

4. True

δ-Aminolaevulinic acid synthetase requires pyridoxal phosphate as a cofactor. In the inherited disease, acute intermittent porphyria, δ-aminolaevulinic acid synthetase appears to be abnormally active producing large amounts of δ-aminolaevulinic acid and porphobilinogen. The symptoms are wine-red urine and neurological disorders. King George III is thought to have had this disease.

5. True

A common pathway of tetrapyrrole biosynthesis from glycine and succinyl CoA exists. The direct precursor of haem in haemoglobin, myoglobin, catalase, peroxidase and the cytochromes is protoporphyrin IX.

Section 22

PLANT BIOCHEMISTRY

22.1 The process of nitrogen fixation in living organisms:

1. occurs only in green plants.
2. occurs only in chlorophyll-containing organisms.
3. is carried out by an iron-molybdenum enzyme complex called 'nitrogenase'.
4. requires the expenditure of 1mol of ATP for every mol of nitrogen (N_2) fixed as ammonia.
5. requires 'reducing power' (six electrons) to reduce 1mol of nitrogen to ammonia.

1. False

Nitrogen fixation only occurs in prokaryotes. Some nitrogen-fixing prokaryotes exist in symbiotic relationship in the root nodules of green plants; many more, however, are free-living.

2. False

All photosynthetic bacteria and most, if not all, blue-green algae fix nitrogen, but so can many other (non-chlorophyll-containing) bacteria such as *Azotobacter*, *Klebsiella* and *Clostridium*, which are aerobic, facultative and anaerobic respectively.

3. True

The nitrogenase complex is a multisubunit iron-molybdenum-sulphur protein of M_r about 220 000. The mechanism of the reaction is poorly understood.

4. False

Nitrogen fixation requires much expenditure of energy whether performed by prokaryotes or (industrially) by humans. The exact number of molecules of ATP required to turn 1 N_2 into 2 NH_3 is not certain, but it is at least 12 and may be as many as 15.

5. True

Probably 2 ATP are required for each electron transferred. Reduced ferredoxin is usually, but not always, the immediate electron donor.

22.2 In photosynthesis in green plants, represented by the overall equation

$$6 \, CO_2 + 12 \, H_2O \longrightarrow C_6H_{12}O_6 + 6 \, O_2 + 6 \, H_2O$$

1. all the oxygen (O_2) evolved comes from carbon dioxide (CO_2).
2. an input of energy is required.
3. an input of 'reducing power' is required.
4. the initial product from the fixation of CO_2 is a 5-carbon sugar.
5. light of two wavelengths is required for optimal efficiency of carbon dioxide fixation.

1. False

All the oxygen evolved comes from the water. This was shown experimentally by substituting the heavy isotope of oxygen, ^{18}O, in the water used. The equation is therefore written with 12 H_2O on the left hand side and 6 H_2O on the right. The overall reaction

is a *reduction* of CO_2 to the level of carbohydrate
(CH_2O).

2. True
Bonds are being formed, since six carbon atoms are
being joined together to produce a hexose sugar;
therefore energy is required. This is provided by
ATP.

3. True
Carbon dioxide is being reduced to carbohydrate
(see 1). The 'reducing power' is provided by NADPH,
the hydrogen of which comes from water.

4. False
The *acceptor* for CO_2 is a 5-carbon sugar (ribulose
1,5-bisphosphate), but the earliest product of carbon
fixation that can be identified is a 3-carbon sugar,
glycerate 3-phosphate (3-phosphoglycerate).

5. True
Green plants have two photosystems each operating
at a different wavelength. Both must be operative
for there to be sufficient energy input to bring about
the reduction of $NADP^+$ to give NADPH.

22.3 In the diagram of a chloroplast, shown below:

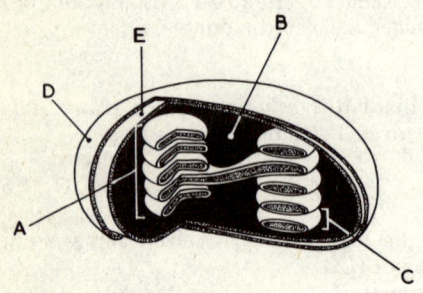

1. A is stroma.
2. B is granum.
3. C is a thylakoid.
4. D is outer membrane.
5. E is endoplasmic reticulum.

1. False
A is a stack of thylakoids, termed a granum.

2. False
B is the stroma, the space within the inner membrane

around the thylakoids and which contains soluble enzymes.

3. True
Thylakoids are flattened sacs which carry out the 'light phase' of photosynthesis.

4. True
Like mitochondria, chloroplasts have an outer and an inner membrane. The outer membrane in each organelle is highly permeable to small molecules and ions.

5. False
Chloroplasts do not possess a structure resembling the endoplasmic reticulum, although they do have free ribosomes and synthesize some protein.

22.4 Although photosynthesis occurs in a number of distinct steps, the overall process can be summed up by a relatively simple equation. Which of the following equations summarizes the processes of photosynthesis either in green plants or bacteria? In the equations, (CH_2O) represents carbohydrate.

1. $CO_2 + H_2O \rightarrow (CH_2O) + O_2$
2. $CO_2 + 2H_2O \rightarrow (CH_2O) + O_2 + H_2O$
3. $CO_2 + H_2S \rightarrow (CH_2O) + S + \frac{1}{2}O_2$
4. $CO_2 + 2H_2S \rightarrow (CH_2O) + 2S + H_2O$
5. $CO_2 + 2\ lactate \rightarrow (CH_2O) + 2\ pyruvate + H_2O$

1. False
See 2.

2. True
This equation describes photosynthesis in green plants where the hydrogen donor is water and oxygen is released. The presence of water on *both* sides of the equation is important. See 5.

3. False
See 4.

4. True
This equation describes one type of bacterial photosynthesis where the hydrogen donor is H_2S and sulphur is produced.

5. True
This equation also describes one type of bacterial photosynthesis. Comparing 2, 4 and 5, a general

equation for all types of photosynthesis can be written:

$$CO_2 + 2H_2A \rightarrow (CH_2O) + 2A + H_2O.$$

22.5 Which of the following take part in the 'light phase' of photosynthesis?

1. Ferredoxin.
2. Cytochrome P_{450}.
3. Plastoquinone.
4. Plastocyanin.
5. Haemocyanin.

1. True
Ferredoxin is an iron-sulphur protein which carries electrons for the reduction of $NADP^+$ to NADPH.

2. False
Cytochromes *are* involved in photosynthesis, e.g. cyt b_{563} and cyt c_{552}, but not cyt P_{450}.

3. True
Plastoquinone is tightly bound in photosystem II, and closely resembles ubiquinone, the quinone of the electron transport chain in mitochondria.

4. True
Plastocyanin is a small, copper-containing protein. It acts as the final electron carrier from photosystem II to photosystem I.

5. False
Haemocyanin is a copper-containing respiratory protein found in the blood of certain invertebrate animals.

22.6 Chloroplasts generate:

1. NADH in light.
2. NADPH in light.
3. $FADH_2$ in light.
4. ATP in light.
5. ATP in darkness if they have first been exposed to a buffer of pH 4 for several hours.

1. False
NADH is not involved in photosynthesis.

2. True
Photosynthesis generates reducing power in the light in the form of NADPH.

3. False
Flavins are involved in photosynthesis (e.g. FAD is the prosthetic group of ferredoxin-$NADP^+$ reductase) but $FADH_2$ is never released or 'generated' as a final product.

4. True
In the light, chloroplasts generate ATP in addition to NADPH.

5. True
Pre-incubation of chloroplasts at acidic pH was the classical experiment which showed that ATP production occurs as a result of the generation and dissipation of a proton gradient.

22.7 The first identifiable product of CO_2 fixation in photosynthesis is:

1. malonyl CoA.
2. ribulose 1,5-bisphosphate.
3. glycerate 3-phosphate (3-phosphoglycerate).
4. glyceraldehyde 3-phosphate.
5. glycerate 1,3-bisphosphate.

1. False
Malonyl CoA is formed as a result of CO_2 fixation in fatty acid biosynthesis but is not involved in photosynthetic CO_2 fixation.

2. False
Ribulose bisphosphate is the immediate *acceptor* of CO_2 in photosynthesis.

3. True
The combination of CO_2 and the C_5 sugar, ribulose 1,5-bisphosphate probably yields a transient C_6 intermediate which immediately breaks down to produce two molecules of glycerate 3-phosphate.

4. False
In subsequent steps, glyceraldehyde 3-phosphate is produced by reduction of glycerate 1,3-bisphosphate (see 5) using NADPH.

5. False
In subsequent steps, glycerate 3-phosphate is phosphorylated using ATP to produce glycerate 1,3-bisphosphate (see 4).

22.8 The operation of the 'C_4 pathway' of photosynthesis in certain tropical plants:

1. allows efficient photosynthesis without the need for ATP.
2. achieves high rates of photosynthesis, even with the stomata closed, preventing excessive water loss.
3. achieves maximum rates of photosynthesis with the stomata open in high light intensity.
4. produces oxaloacetate, which is quantitatively converted into carbohydrate.
5. allows photosynthesis to continue in the hours of darkness.

1. False
The C_4 pathway (Hatch-Slack pathway) of photosynthesis actually uses more ATP than does the C_3 pathway alone.

2. True
In the tropics, light intensities and temperatures are high but the humidity may be low. Complete opening of stomata would lead to excessive water loss by transpiration. The C_4 pathway allows CO_2 to be fixed more efficiently at low concentrations (i.e. with the stomata almost closed). This is because the first enzyme in the pathway, phosphoenolpyruvate carboxylase, has a much higher affinity for carbon dioxide than carboxydismutase of the C_3 pathway.

3. False
See 2.

4. False
Although oxaloacetate is produced in this pathway, it is immediately reduced to malate in the mesophyll cells. In the bundle-sheath cells, the malate is converted to pyruvate with the release of CO_2 which then takes part in the ribulose bisphosphate pathway of CO_2 fixation.

5. False
The primary event of all photosynthesis is the conversion of light energy into usable chemical energy. Therefore no photosynthesis can take place in the dark.

22.9 Plants use chlorophylls during photosynthesis:

1. as light-absorbing screens to prevent damage to the sensitive photosynthetic apparatus in the thylakoids.
2. as hydrogen donors in the production of NADPH.
3. to make the thylakoid membrane more permeable so that a gradient of protons can be generated.
4. as photo-oxidizable compounds capable of supplying electrons to ferredoxin and eventually to $NADP^+$.
5. as acceptors for electrons derived from CO_2.

1. False
The absorption of light by chlorophyll molecules represents the first step in the conversion of light energy to chemical energy.

2. False
See 4 and 5.

3. False
Although a gradient of protons is generated, this results from electron flow through other carriers such as quinones and plastocyanin. Membrane permeability is not affected by chlorophylls.

4. True
The action of light on a chlorophyll molecule 'ejects' an electron, leaving the chlorophyll in a more oxidized state. The ejected electron passes to ferredoxin and can ultimately cause the reduction of $NADP^+$. It is replaced in the chlorophyll molecule by electrons taken from water in green plants or from other, oxidizable compounds (H_2A) in photosynthetic bacteria.

5. False
Water (or H_2A) supplies the electrons (see 4). Carbon dioxide eventually acts as *acceptor* for electrons (and protons) when it is reduced to the level of carbohydrate. However, CO_2 does not participate in the stages of photosynthesis that involve chlorophyll.

22.10 Which of the following features are exhibited by both green plant photosynthesis and bacterial photosynthesis?

1. The presence of two photosystems.
2. The utilization of H_2O as electron donor.
3. The possession of chloroplasts.

4. The utilization of some form of chlorophyll as light receptor.
5. The production of ATP, utilizing the energy stored in a gradient of protons.

1. False
Green plants have two photosystems whereas photosynthetic bacteria have only one.

2. False
Only green plants use water as the reductant during photosynthesis.

3. False
Only green plants have chloroplasts. Bacteria have chromatophores which are analogous in function but are quite different in structure.

4. True
It seems that in all forms of photosynthesis, chlorophyll is involved. In green plants, it is chlorophylls a and b; in bacteria, it is bacteriochlorophyll.

5. True
The mechanism of ATP production during photosynthesis is very similar to that occurring in mitochondria.

INDEX

(* Indicates a question on the topic)

243

248

Acknowledgements

We are grateful to Anne Turner for efficient prepara-
tion of the manuscript and to colleagues in the
Department of Biochemistry, University of Leeds and
elsewhere for helpful advice and comments.